HOW TO BE A *GOOD* ENGINEER

SANTIAGO BARCÓN

HOW TO BE A *GOOD* ENGINEER

TEN COMMANDMENTS AND 100 TOP TIPS

LID

MADRID | MEXICO CITY | BUENOS AIRES | BOGOTA
LONDON | SHANGHAI

Published by:
LID Editorial Mexicana S. A. de C. V.
Homero 109, 1404, Colonia Chapultepec Morales,
Mexico, Mexico City, 11570
Tel. +52 (55) 5255-4883
mexico@lidbusinessmedia.com - LIDlibros.com

A member of:

BPR

businesspublishersroundtable.com

Original title: «Cómo ser un buen ingeniero. 10 mandamientos y 100 consejos para lograrlo».

Translated by author.

ISBN paperback: 978-607-26988-3-3
ISBN *e-book*: 978-607-26988-4-0

Editing and proofreading: Gail Friedman
Cover Design: Juan Ramón Batista Pérez
Cover Layout: Javier Perea
Typesetting: www.produccioneditorial.com

First edition: June 2025

Acknowledgments

In a professional career that has already spanned four decades, it is impossible to include everyone to whom I owe what I have achieved. Nevertheless, I begin with my family: grandparents, parents, siblings, wife, and children. And to my teachers, undoubtedly apostles of education, and especially those who really challenged me.

To my engineering colleagues with whom I fortunately maintain close contact. I have learned from hundreds of you over the course of my professional life, but I would particularly like to highlight Johann Inge Westman, and my partners: Rodolfo Maza, Genaro Maza, and Ramón Delgadillo (the first two of whom have unfortunately passed), for having absolute faith in the projects I presented to them.

To Rafael Guerrero, with whom I have co-authored another book, for writing the prologue to this one; to Eduardo Marchesi and Mario González, for their recommendations on the back cover and their valuable friendship.

To Thania Rodríguez, who supported me in transcribing the manuscript into electronic format. Also, of course, to Ada Laura Luna, Claudia Herrán, Norma Ramos, and Bertha

Herrerías, from Editorial LID, who guided me in bringing this work to fruition.

Finally, for this English Edition, thanks to my great friend from New Zealand, Geoff Vaughan who, together with his wife Judy and their daughter-in-law, Brie Jessen-Vaughan, reviewed the English translation and offered valuable advice.

Dedication

To my family: wife, mother, children, daughter-in-law, son-in-law, and brothers, and especially to Almudena, Fátima, Lorea, Lucía, Marina and Santiago, my grandchildren. I hope that some of you will decide to follow this exciting profession.

Also, to all young engineers and to those who are already experienced, may you obtain benefits that will enable you, as the title of this work suggests, to be good engineers.

Contents

Competences and values

Personal and professional development

Professional practice

Foreword

After reading the book, *How to be a Good Engineer. Ten Commandments and 100 Top Tips,* I recognized myself in several pages of the text. During my student years, I would have «fallen in love» with an orientation of this kind, or with the accumulation of advice contained in these pages.

When dedicating a few words, it is difficult to stop the hand that, left alone, would tend to praise the dear friend. Aware of this risk, I therefore resort to the following anecdote in order to speak about the author. In the seventh century B.C., a blind chronicler and poet narrated a Mycenaean legend that had occurred in the thirteenth century B.C. In it appears the great master craftsman of the Bronze Age, forerunner to current knowledge of electrical engineering, control, artificial intelligence, and transient phenomena, among others. He designed and built a series of artifacts, all of which were useful. This budding engineer would have made Hermes's winged sandals and helmet, Achilles's armor, the bow and arrows of Eros, the crotalum of Hercules and, as if that were not enough, Agamemnon's scepter and the thunderbolts of Zeus.

Perhaps most interestingly, from the point of view of robotics and artificial intelligence, he designed and built two golden maidens (*κομραι χρμσεαι*), two automatons, with the extraordinary appearance of young women. These not only served the engineer and his guests with sweetness and loving care but also sang with beautiful voices.

This is how I see Santiago Barcón: tenacious, always looking for the best solutions, directing an accredited magazine and navigating the rough and turbulent sea of industrial competition, where only the fittest evolve and prevail.

Back to the text. The book, *How to be a Good Engineer. Ten Commandments and 100 Top Tips* assumes that the reader, a sure beneficiary of its contents, has finished their degree. Because of this, it presents a timely and impeccable decalogue for engineers. The precepts are organized into three sections - Competences and Values, Personal and Professional Development, and Professional Practice - which, in turn, are broken down into «Knowledge Capsules»- small doses of knowledge about the importance of a holistic engineering education.

Reading invokes memories. In the engineering school where I obtained my degree as an electrical mechanical engineer, I do not once remember seeing an advisor, guide, or tutor walking the halls; everything was ultimately left to the responsibility and risk of the student. If a class was poor, regular or just deficient, it was up to the students to complement it. I nevertheless have fond memories of my thermodynamics, algebra and geometry teachers, while I was forced to buy a book and study on my own for my key subject, Electrical Machines.

In an informal chat with Santiago over a cup of good coffee, he told me he is preparing a second volume. It will deal in a similar way with what happens during the period between entering the faculty and graduating with a professional degree. He also said that the volume would place importance on the ever-expanding role of the mentor, as each can tutor between twenty and forty students. This is no small matter if one considers that the training provided by this figure confers great professional standing, aimed at the success of their students.

This book is an enjoyable and fascinating read, full of real-life examples, inspiring quotes, and practical advice derived from the author's own experiences. Its approach offers a new look at the role of engineers in their field of expertise. It is an undoubtedly interesting one for engineers of all levels, from students to established professionals.

Rafael Guerrero Cepeda
Engineer, Former Head of the CFE
Advanced Engineering Unit

Introduction

It is surprising that the syllabus for an Electrical Engineering degree, my degree, does not include a history of our profession as a compulsory subject. The same applies to other disciplines. Let us bear in mind that at the end of the 20th century, the Academy of Sciences of the United States of America considered that the most important change of the previous century had been the construction of an interconnected electrical system which, as we electrical engineers like to boast, is the greatest machine ever built by humans.

Reviewing the history of engineering not only increases general knowledge about the late nineteenth and twentieth centuries but also allows us to understand the challenges faced and how some technologies prevailed despite not being the best. It would also show us how the ambition of some capitalists unfortunately deprived the true creators of certain innovations of the time of the subsequent economic benefits. As an example, the radio was patented by Tesla, but Marconi, together with his associates, usurped it. Combining the theft with impudence, they returned the license to Tesla on the day it was due to expire.

On the other hand, it is surprising that names such as Charles Proteus Steinmetz, Edwin Harder, and others remain unknown, ignored by colleagues who continued their work.

I firmly believe that a concept is easier to understand if you know where it comes from, hence my insistence on studying the origins of engineering; for if we do so, we will surely become more deeply attached to our profession.

It is necessary to read -even in *Wikipedia*- about little recognized «giants»: Jean Marie Ampère, William B. Shockley, John Bardeen and Walter H. Brattain, James Clerk Maxwell, Heinrich Hertz, Alessandro Volta, Benjamin Franklin, Zénobe Théophile Gramme, Charles Brush, William Stanley Jr., Thomas Murray, Frank Sprague, and Lee De Forest. Not everything involves Edison or Tesla.

Knowing and understanding the history of our profession allows us to better understand engineering, an activity that began 400 years ago in France. Its birthplace should not surprise us as the etymology of the word 'engineer' is neither Greek nor Latin, but much more recent. At that time, artifacts were called engines and were complex mechanisms that showed intelligence in their design.

In its beginnings, engineering was linked to military issues and needs. In 1676, the first engineers, Frenchmen, who formed part of the army infantry, were organized into special units called *corps du génie*. These units were both a center of development and of engineering. They were also places where «engineers» were trained, referred to as *officeurs du génie.*

Less than two decades later, these groups became famous throughout Europe for their achievements in military-related constructions. Similar to what we see today, the groups were structured in an organization of masters and apprentices, however, without a clear academic syllabus, or anything close.

Other countries, anxious not to be left behind, borrowed the French word, *ingénieur,* to define this type of person, and so it has continued to the present day. By the end of the eighteenth century - about ninety years later - a study program had been developed that is very similar to the one we know today.

Its military origins will forever mark this profession: reliability, speed, and ease of use for a specific purpose have always been priorities. Given that the training of the first engineers was essentially military, discipline in practicing the profession is an enduring legacy. In addition to these characteristics, there was also preparation in mathematics and physics.

All this professional orientation was assumed by the *École de Travaux Publics* created in 1794, becoming the *École Polytechnique* the following year. Its three-year academic curriculum established the guidelines that we continue to follow to this day: theory and then practice (the similarities between the curricula of the *École de Travaux Publics* and the *École Polytechnique* of the time and current ones are immediately recognizable).

This brief history is completed by the exponential growth of this profession. In 1850, the first census in the United States that recorded the number of engineers reported a total of 2000, in a population of 23 million, that is, about one for every 10,000 people. By 1980, the ratio was 1/100, i.e., it had increased 1000 times.

I invite you to delve into the history of engineering. Not only is it a fascinating subject, but it also encourages us to remember the concepts we studied in our youth. I assure you that historical knowledge of our profession is what can differentiate us from our colleagues and make us more reliable in the eyes of our clients.

Let me give you an example: when I tell clients that the first rule of limiting harmonics was telephonic interference, as electricity and communication conductors were on the same circuit, they (the client or boss) realize that I am not «just another engineer,» and I also demonstrate something that is contagious: a passion for engineering.

The Ten Commandments of an Engineer

I. Love your profession above all other activities.

II. Do not minimize the decisions you have to make.

III. Take breaks that allow you to be productive and safe.

IV. Honor your colleagues and competitors.

V. Do not put others at risk.

VI. Give back to society more than you have received.

VII. Do not steal or be corrupt.

VIII. Do not lie.

IX. Do not base your professional success on financial achievement.

X. Continue to learn until the last day of your professional life.

I. Love your profession above all other activities.

As an engineer, you should have an absolute passion for your profession. It should be the focal point of your work activities, and you should not prioritize any other activity above it.

If you do not keep this high level in mind, the degradation of your performance will become more evident day by day.

Love also means pride, so you should defend engineering with the same zeal as you would your family.

II. Do not minimize the decisions you have to make.

Being an engineer implies having a clear and firm vision of what you have in your hands. The safety and well-being of your fellow men and women depends, first and foremost, on your decisions and recommendations. Remember that omission is itself a sin, and perhaps the worst of all.

III. Take breaks that allow you to be productive and safe.

The mind must be clear to be productive. There is no way to develop your full potential if you are not fulfilled. This is even more relevant when in contact with equipment and machinery.

Just as airplane pilots have mandatory rest hours for each period of activity, you should consider the same for yourself.

IV. Honor your colleagues and competitors.

Respecting your colleagues and competitors is vital. Therefore, you must keep a cool head and an open mind to their comments and contributions.

True, there are some who will not deserve it, but they are the minority. Resorting to generalized disparagements is a serious mistake and a sign of a lack of basic intelligence.

V. Do not put others at risk.

It is essential that the first issue to verify, in all your decisions and actions, is safety. Latin Americans tend to be lax in this area, with fatal consequences.

There is nothing more precious than human life: to put it at risk is unacceptable. I would add that, if in doubt, consult or abort. The «I thought that...» is not acceptable. Again, safety always comes first.

VI. Give back to society more than you have received.

You have a moral and personal obligation to return to society more than it has given you. We tend to think that we have earned everything by the sweat of our brow, but there is not a single example where this is the case, regardless of whether you studied in a public institution, paid for by taxes, or in a private one, where you are part of a privileged minority.

There are hundreds of ways to give back: give classes, participate in forums, write, provide guidance, listen to those who ask for advice and so many more. If you do not intend to make these a habit, or if your only goal is to simply «maintain the status quo,» you will certainly never make a significant contribution. You have to create a multiplier effect.

VII. Do not steal or be corrupt.

Theft is not only taking ownership of someone else's property; it includes delivering a product out of specification or writing a report without engineering rigor; it is also producing a design that only just complies with the lower limit (millimeters away from non-compliance).

Giving «financial compensation» to customers to accept a poor service or product is unacceptable: it is the famous Mexican «*mordida*», the Chilean «*a la cochiguagua*», the Colombian «*mermelada*», the American «kickback» or whatever other nickname by which it is known. Hide your head in shame if you even consider it.

VIII. Do not lie.

It is sad to see pseudo-engineers who say: «My harmonic filter is Mexican technology and manufactured in Mexico,» when, in reality, they are Chinese components that are only assembled in Mexico. In addition, many even have the audacity to say that their components have artificial intelligence.

Using nationalism to sell is crude. Moreover, any engineer who plays along with the boss or owner of a company is, to say the least, embarrassing and a disgrace to the profession.

Remember that a true engineer does not take shortcuts or distort reality for their convenience nor to fulfill the whims of their superiors. Falsifying is the same as lying, make no mistake.

Solutions that are not robust, that do not have a long-term vision, or that fail to think about the user are unacceptable. The same applies if the place where they will be installed and, of course, the safety of the operation is not considered.

«The truth will set you free» is a phrase attributed to Jesus Christ in the Gospel of St. John. You cannot lie and, moreover, you must point out or denounce when someone else does. Remember the Mexican saying «he who kills the cow sins as much as he who ties its leg.» Do not look for petty excuses to justify your lack of integrity or the cowardice of turning the other way.

IX. Do not base your professional success on financial achievement.

It is sad to see engineers judge their peers by the amount of money they have made, without even asking where it came from.

Sadly, in most cases, that money does not come from their engineering expertise, but from their business skills (at best), and can be seen in the poverty of their work and installations: from buildings that collapse in earthquakes, electrical failures that cause fires and fatalities, and hundreds of other deficiencies. They lie outright to customers to make significant profits. Lying is something they use to justify their actions to themselves in order to achieve their goal.

We must respect each other and give appropriate recognition to those who are really good engineers.

X. Continue to learn until the last day of your professional life.

The world is changing rapidly, but even more so in our daily work. An engineer who does not keep learning, updating their skills, studying, interacting with colleagues, is just a museum piece. Without this discipline it is impossible to embrace innovation and keep curiosity alive, the cornerstone of knowledge.

Belonging to societies or associations is essential, take courses, consult and, most importantly, share knowledge. I quote Arn Anderson: «I am both a teacher and a student, if you are no longer a student you have no right to call yourself a teacher.»

If you lose your curiosity and appetite for learning, you are doomed to remain outdated.

100 Top tips to be a good engineer

compet
and val

ences

ues

COMPETENCE: the ability to perform something skillfully. It implies not only the performance of a task, but also its execution, taking account of its importance and finding the optimal way to carrying it out.

VALUE: desirable attribute in specific fields of society and culture.

If you have integrity nothing else matters. If you do not have integrity, nothing else matters.

Alan K. Simpson

01

PERSEVERANCE

Perhaps the attribute that contributes most to a person's success is perseverance. You should not «throw in the towel» at the first setback. You should know how to deal with the bad times; trust that there is always light at the end of the tunnel; and remember that you have solved your problems and overcome adversities in the past.

Without this conviction, projects can become truncated and, even worse, can generate a feeling of helplessness, of not being able to achieve what you set out to do, which in turn creates a downward spiral.

Success stories will often highlight a triumph, however people do not fully appreciate the enormous effort involved in obtaining such triumphs. That is why when you tell a successful person that they got lucky, their response is usually explosive and indignant.

Thomas A. Edison said: «Genius is 1% inspiration, 99% perspiration»; Goethe, for his part, is credited with the maxim: «Without haste, but without pause, like the stars,» and let's not forget Uncle Scrooge who (leaving aside the issue of his greed) perseveres until he obtains what he wants: he never gives up. In my opinion, success is work plus perseverance.

Perseverance is not usually a quality that distinguishes Mexicans, as we tend to get discouraged and assume a defeatist attitude that we often compensate with optimism: «We'll get it next time.»

Do not fall into this trap. It is literally, «the kiss of death» for a full life.

02

SPEAK ENGLISH (OR ANY OTHER SECOND LANGUAGE)

A 21st century engineer cannot afford to not speak English or to be monolingual. I do not mean the ability to stammer a few words or read it, to get by with 60%, 70% or 80% of the language. We must master it and be able to maintain a fluent conversation on different topics, without hesitation. We should be able to understand different accents and, most importantly, overcome the false shame of making mistakes. We should remember that our English is often far superior to the Spanish spoken by most Americans or English.

With a sense of false nationalism, some may think that learning another language is tantamount to «selling out our homeland.» But nothing could be further from the truth. What does, without doubt, betray our origins, however, is speaking and writing Spanish incorrectly.

But back to Shakespeare's language. The vast majority of articles, papers, and other current sources are in English. Without access to this information, you will be hopelessly unable to thrive or, at the very least, keep up to date.

So, let's be clear, if English is your mother language you may not need to speak another, but it will give you a great competitive advantage.

I am fortunate enough to have studied English since I was a child, and when I finished my degree, I received an excellent job offer from the multinational company, 3M. However, I had already accepted a position in the Mexican company, Vitro. When negotiating my salary with this company from Monterrey, I highlighted that I spoke English, to which they responded: «We are a Mexican company, we do not require English.»

Hardly three months had passed when I became the interpreter for foreigners visiting the plant. Needless to say, in addition to being in permanent contact with the company's top management, I always learned first-hand from experts in various subjects. Additionally, I was sent for training in different cities such as Toledo, Ohio, and Boston for six months. It was a highly valuable experience.

Learn English (or any other language) or do not blame the company or luck for overlooking you for promotions.

I end by quoting my great-grandmother, Lola, from more than fifty years ago: «In these times you have to know how to speak English and how to swim.»

I conclude by saying that if I had to choose between my master's degree in management or speaking English, without a doubt, I would choose to master the language of Shakespeare.

03

TOLERANCE

Engineers deal with exact values that are verifiable. This, coupled with our mental makeup, can make us a little intolerant.

I remember that once, in a meeting to analyze a large-scale project, the chief engineer, who was an extremely capable person, said some nonsense about the operation of the lightning rods. At once, several of the participants who were almost thirty years younger than him, corrected him, proud of having shown him up.

As expected, the boss reacted by sticking to his opinion and the discussion escalated. I remained silent. Suddenly, he blurted out, «You, who probably knows, tell me, am I right or wrong?» I replied that he was right, and we moved on to another topic.

At the end of the meeting, the younger associates approached me to complain. I explained that since the position of the lightning rod decreased the connection to the grounding system by ten centimeters, it did indeed have a lower impedance, just as the boss had said. Although it was infinitesimal, it was, in fact, true. Of course, the value I was referring to was totally negligible, but it is all a matter of how you look at things.

A few days later, I met alone with the chief engineer and commented: «I have been studying up on things, and the effect you referred to in the meeting is insignificant. In general, we shouldn't even consider it.» In another meeting that same afternoon, the head engineer immediately recognized his «mistake» and discussed it with those with whom he had previously disagreed.

Knowing how to accept mistakes is the hallmark of great engineers, and also the only way to learn and move forward.

04

DO NOT CREATE IDOLS

Worshipping or paying obeisance to idols can lead to some very serious mistakes. Revering something or someone is understandable, but although it may be intended to protect our image and beliefs, it often leads to mistakes.

The «best» brands or the ideal technological solutions can easily become idols, but sooner or later, they may crumble. That is why studying the history of our profession is so helpful. We have already seen Westinghouse, AEG, Sprague, Cooper, Alstom and thousands of other companies disappear. Even General Electric is in a complicated situation today, when in 2000 it was the most valuable company in the world, not to mention that many others, mainly European and Chinese, survive only because of subsidies from their governments.

I predict that Elon Musk will soon be part of this list. In short: everything should be held from a proper perspective and do not doubt that sooner or later, all idols fall.

05

EXPERIENCE

«The devil knows more from being old than from being the devil» is a saying that we have heard since we were children and that has become true after so much repetition.

Unfortunately, experience is not directly proportional to age, nor should these be confused.

When I see a resume where candidates state that they have 17 years of experience, I always ask myself: 17 years of experience or one year repeated 17 times? The difference is crucial, and we should not make this mistake.

Experience means learning new things every day and broadening our field of knowledge to take on new ideas, recognize good and bad practices, identify what needs to be taken care of in our projects, investigate new solutions, and dozens of other things.

To do so, we need to get out of our comfort zone, accept new challenges and, of course, never stop learning. Experience is not the same as length of service. Opt for the former and stay away from the latter.

06

PROBITY AND THE COURAGE TO REPORT

We live in a society that, according to international evaluations, has high levels of corruption and thus suffers from a lack of probity.

If you learn something new every day, you can teach something new every day.

Martha Stewart

We tend to associate this phenomenon with kickbacks and bribes, but in reality, these are just one facet. Not fulfilling commitments, taking «shortcuts,» or any thought of «shortcuts,» all fall into the category of «after all, it doesn't matter that much!» But they are equally serious, though not as obvious, breaches of probity.

Honesty, the twin sister of probity, implies not taking money or goods that do not belong to us, but also means not wasting resources, squandering time, or doing work half-heartedly, among other things. Integrity is also a sister and, as in a Venn diagram, these all intersect.

There is nothing more important than being honest. Without it, there is no decency and without that, we cannot go through the world with our head held high. This training can be reinforced at school or university, but really, it is at home where it is acquired. This kind of courage is instilled in us from birth. It does not come from a course or from attending a weekend religious retreat; that is why it is so difficult to find and so valuable at the same time.

For example, an honest engineer will not put in a conductor gauge that is cheaper than what was quoted, even if it is surplus to the desired performance. And if they did, they would adjust the cost - what was saved - on the invoice. The same is true if a different cement or steel is used than what was quoted.

There are thousands of examples, and we all know what the right thing to do is: it basically consists of not thinking that we are smarter than the rest of the world, as well as not succumbing to «everyone else is doing it.»

On the other hand, an engineer must be particularly courageous to denounce what is wrong and defend just causes. And not just because it is the ethical thing to do, but because in this way we earn the respect and recognition of our colleagues, even if that goes against the boss.

In general, we do not stick our necks out because it may be easier or more comfortable to turn a blind eye or because we ask ourselves, «What do I have to gain by getting involved?» Probably nothing, but it is a fact that self-respect is lost. It is a downward spiral: we may think, «If I was able to accept that, then this is not so bad.» In the end, everything seems acceptable, and we lose our point of reference.

We must denounce corruption, the colleague who does nothing and who, in addition, complicates the lives of those who do want to grow and contribute. It is our professional and moral obligation. It involves exposing moral or economic wrongdoings, the abuses that have been carried out and, above all, those who hide behind «legality.» When people say: «It is not illegal,» I hear: «It is not ethical.» In short, it has to do with any action that can harm someone.

When denouncing something, do not use euphemisms or sugar-coat your words on the pretext of not hurting someone's feelings. A typical example can be found in the Federal Electricity Commission (CFE), and in many other electric companies where what is simple theft is sometimes referred to as «non-technical losses.»

Another clear example can be seen in the debate that occurred in the electricity sector as a result of the proposed changes to the Energy Reform in Mexico: I defend my interests and use any argument to benefit myself, even if it harms the country; even worse, I do not offer alternatives.

Civil courage entails not minimizing the facts because, in the end, this only results in short-term satisfaction. «An aspirin to fight cancer.»

There will always be excuses or creative ways to rationalize these acts, such as «I need the job» or even more cynically, «it's the same everywhere.»

Corrupt people exist and always will exist, but their number can and should be limited by the only efficient mechanism: denouncing.

I have denounced corruption on several occasions, and yes, it is not easy. Even worse, the «mafia» tends to close ranks and complicate the process for securing orders in the future. Nevertheless, over the years this is corrected, and bad practices are eliminated, at least in our sector. A colleague in the industry once asked me, not without a hint of derision, if I could end corruption in the company we were working with. My answer was simple: «Of course not, I only watch and act where I do business, the others can take care of their own business.»

But I am convinced that if we act in the right way, corruption levels in our country would be much lower. So, for our own sakes, but also for Mexico, we must speak out.

07

DIVERSITY AMONG ENGINEERING

Jokes among colleagues are always part of student and professional life; but jokes aside, they are also a source of learning and an opportunity to take advantage of diversity. An electrical system can be explained using mechanical analogies. This allows us to see what is needed from another area and to understand that every profession can benefit from another.

Throughout my career, I have been surprised to find fellow engineers from other specialties who have a better understanding of an electrical power system than electrical engineers themselves.

This, of course, has relevance in electronics, as power electronics has advanced significantly in recent years and their use in distribution and transmission is becoming more common. Active harmonic filters, static var compensators, STATCOM, and dozens of other products are already in daily use and have found their niche. The same applies to building materials, chemicals and so much more.

Therefore, a good engineer should not limit themself to a single discipline, but be open to and learn from related areas.

08

HUMILITY

The best way to recognize an excellent engineer is, without a doubt, when their attitude and deeds show humility. This, of course, includes always speaking in the plural, but it is only a beginning.

The acid test is when a mistake is made or something is not known. Knowing how to say: «I made a mistake» or «I don't know» is the best measure. In fact, the only valid one.

Humble behavior, in addition to facilitating knowledge, allows for closer and more productive relationships with colleagues: no one likes to deal with a proud but ignorant person.

09

DO NOT LIE

An engineer does not have to obey all of Moses's commandments, but the one they must obey, without fail, is the eighth: «Thou shalt not bear false witness or lie,» especially when it comes to technical specifications or product performance.

Lying about the characteristics of a product can cause very serious problems and even violate the fifth commandment: «Thou shalt not kill.»

Equipment that will not work as promised can cause personal misfortunes, either directly to the operator, due to the consequences of the failure, such as a possible fire, or, as we sadly experienced in Mexico City and other seismic zones, with collapsing structures.

There is no room for half-truths or thinking that «it's good enough.» Trying to minimize the consequences of poor work only makes things worse and exponentially increases the risk.

This comes from an idea that resembles a very common excuse: it meets the standard.

Remember that the standard is the «minimum of minimums,» and that you have the responsibility and obligation to match the product or design to the task that it will perform.

If you cannot avoid lying or cheating, rather engage in another activity where dishonesty does not cause physical risks to others. You may get away with selling 900 milliliters of gasoline as a liter, or 800 grams of sugar as a kilo, but never say that a piece of equipment will withstand a short circuit, or that it will not overheat, or that a structure can resist 8.1 quakes on the Richter scale. All these examples are clearly untrue, but there are different magnitudes, and the consequences are very different.

10

DO NOT ADULTERATE THE PRODUCT

Modifying the nature of a product is a serious offense, but even more so if this compromises health and physical well-being. Examples include watered down alcoholic beverages, expired medicines, or placebos, but similarly, construction materials that do not have the specified strength, chemical products with reduced active content, and electrical equipment that, when it fails, emits toxic gases.

This adulteration takes several forms: conductors with lower ampacity, poor lighting equipment without safety testing, switches without the required interrupting capacity, and metal clad panels that are only able to pass prototype tests by cheating.

Undoubtedly, we have all faced this scourge and there are only two ways to combat it. The first, simple, is to keep a close eye on what you buy. The second requires courage: report it.

The only way to prevent this fraudulent conduct from increasing, and perhaps even to diminish it, is to file formal complaints with the authorities and proceed legally. Otherwise, we are co-conspirators.

11

«WE» NOT «I»

A good engineer must learn, from day one, that it is essential to speak in the first-person plural: «We» and avoid like the plague the first-person singular: «I.»

If there is one thing that destroys cooperation, it is egomaniacs; people who, in their eagerness to appear important, even indispensable, forget everyone around them. This naturally results in the disappearance of the support of their peers at the slightest stumble. Speaking in the plural to include everyone may seem simple, but it requires a great deal of discipline and self-confidence.

12

INVEST IN YOURSELF THROUGHOUT YOUR LIFE

We are already engineers. Now we must be the best we can be, with all our effort and will. No one, no matter how smart they are, is able to reach their full potential unless they continue to study, attend conferences, read and, importantly, write.

Another basic and perhaps indispensable tool is teaching, educating others. Dealing with students (many of whom maintain that spark) forces us to continue to upskill and to correct our mistakes.

Membership of a society, such as the Institute of Electrical and Electronics Engineers (IEEE), the Mexican Institute of Chemical Engineering (IMIQ), the College of Civil Engineers of Mexico (CICM), the College of Mechanical and Electrical Engineers (CIME), or whichever one you choose, is both an investment and an intellectual obligation.

It is surprising how many professionals do not strive to further their own skills unless their company provides them with such training. This is petty, clumsy, and short-sighted. If we think education is expensive, try ignorance. We must open our wallets, and our brains will follow.

13

ABOUT LOYALTY

We all have a great appreciation for loyalty, as it allows us to act with greater freedom and make decisions with peace of

The chief enemy of good decisions is a lack of sufficient perspectives on the problem.

Alain de Bottom

mind, knowing that we are supported and will not be alone; but loyalty has a limit and, to preserve it, you cannot sidestep moral values and security. Never lose sight of the common good. Continuing to participate in a project knowing that a product or service does not meet the specifications, is not right, no matter how much a boss or friend may insist.

It is easy to justify your actions and say, «I was just following orders,» but a good engineer cannot accept being a puppet. It is an insult to colleagues and the profession.

14

SURROUND YOURSELF WITH PEOPLE THAT MAKE YOU BETTER

Unfortunately, mediocrity exerts a tremendous force of attraction. It is often much easier to get along with people whom we surpass in quality, even though this may lead us to stagnate and to walk the path of least resistance that is inexorable.

If we feel too comfortable with our group, it is likely that we need a change in order to continue growing. It is not easy, but there is no alternative if we aim to improve as a person.

15

POLYMATHY

This word, little used or known, describes a person who knows a lot about various subjects. This is also known as a Renaissance Man. Leonardo da Vinci, the clearest example, was a painter, sculptor, engineer, inventor, writer, and was versed in different subjects, as well as speaking a variety of languages.

Polymathy offers people the satisfaction of moving in diverse cycles, making their brains more useful and allowing for the pollination of new ideas. The phrase «old parrot no longer talks» does not apply to this type of person.

Constantly learning something new keeps us alert and up to date; it allows us to meet new people and to have more interesting conversations. Interestingly, it does not necessarily have to be related to what we do; on the contrary, the more varied it is, the more new experiences we are likely to have.

Additionally, some studies on aging, although yet to be proven but certainly logical, have looked at the relationship between constant learning and a significant decrease in the possibility of developing Alzheimer's disease. The brain is a muscle, and as such, we must exercise it.

16

TRAIN AND STUDY ON A PERMANENT BASIS

Benjamin Franklin and Thomas Piketty agree on a crucial point: the importance of investing in education and lifelong learning. Franklin urged us to «empty our pocketbook into our mind» so that it can, in turn, «fill our pocketbook,» thus highlighting the idea that investment in knowledge is fundamental to personal success and to the development of a country. Piketty, for his part, supports this notion by demonstrating the correlation between education and a nation's level of economic development.

Within this context, it is alarming that many engineers, once they have finished their studies, stop looking for new ways to learn. This mentality is at odds with the idea that knowledge is a permanent and invaluable possession. As my grandfather used to say: «The only thing no one can take away from you is what you know.» Lack of continuous updating and training will almost certainly lead to professional mediocrity.

The digital era offers a wide range of accessible educational resources, however the fact that they are free can lead to procrastination and generate a lack of commitment that blocks progress. Let me tell you something: I am a music lover, but I did not play any instrument. I decided to learn one and I opted for the harmonica, which is easy to carry and can be practiced anywhere, even while traveling. It looks like a simple instrument, but it is not. I found hundreds of free online courses to learn to play, but decided to pay for weekly sessions with a teacher (in Uruguay). This forced me to study and practice (another reason for choosing the harmonica is that I work in the field of harmonic filters!)

With this personal story, I want to emphasize that it is essential for new generations to understand that learning does not end with a university degree. As Einstein said: «Once you stop learning, you start dying.» To this must be added the lack of rigor of some educators today, an attitude that limits the potential of students and can harm their future development. So tattoo this on your forearm: a university degree is only the beginning of daily preparation.

17

THE IMPORTANCE OF PRACTICE

«Don't practice until you get it right. Practice until it's impossible to get it wrong.» Andrea Pirlo's phrase teaches us that everything takes assiduity and continuous effort.

Mastering any area of knowledge or activity, such as sports, requires at least 10,000 hours of dedication. Falling short of the energy required to do so will make you only moderately capable, and you will never stand out from your peers.

- Want to write fluently? Practice for 10,000 hours
- Want to have vast general knowledge? Read for 10,000 hours
- Want an enviable physical condition? Exercise for 10,000 hours

These may seem impossible goals, but they are not. The year has 8,760 hours, if you dedicate two hours a day to these activities (or others), in less than fourteen years you will have achieved them. If you are still young, with today's life expectancy, you have about sixty years left. Start today, otherwise, you will never do it.

It doesn't matter if a cat is black or white as long as it catches mice.

Deng Xiaoping

18

WRITE

There is nothing more important than writing. Without this skill, our ideas remain in the void that exists when the sound of our speech is extinguished. No matter how hard we try to express our opinion or contribute, if we only do this orally, sooner or later, it will be lost. This is why writing is so imperative. I know that the fear of the blank page considerably exceeds the fear of standing on the edge of a cliff. Even great writers struggle to begin their work.

Perhaps one of the most regrettable losses in educational institutions in recent years has been the ability to communicate in written form. This is even more evident in engineers as we have always underestimated the ability to communicate in writing. We take for granted that «numbers are enough for us,» which is a serious mistake. Eliminating the thesis requirement for our degree has exacerbated this weakness; perhaps we would not have made a major discovery, but it certainly forced us to write, to give structure to our ideas, and to present them in a coherent manner. Of course, those of us who had to do it hated it, but in retrospect it was worth it. Writing well in engineering simply requires a certain order: check your spelling, strive for acceptable syntax, give yourself a few minutes to proofread your texts. The rest will come with practice.

If you are careful with the wording of an email or a WhatsApp reply, you have already taken the first step. Peter Drucker, one of the greatest thinkers of the 20th century in both the business world and beyond, used to say: «The one basic skill is the ability to organize and express ideas in writing and speaking.»

Two pieces of advice: First, read aloud what you have written in order to pay attention to how it sounds, and, of course, to notice where a comma or a semicolon is needed. Second, do not assume that Word's proofreader is infallible. If the text is important, it is necessary to review it and read it from the end to the beginning; yes, in reverse. That way you will be able to observe word by word and your brain will not fill in the gaps or errors.

Having said all this, I close with a sad and worrying statistic: less than 2% of engineers write.

19

THE VALUE OF READING AND ORAL EXPRESSION

Harry Truman, 20th century U.S. president, once said that not all readers are leaders, but all leaders, at least the real ones, are readers. One should not read to appear to be a scholar, but to be better every day.

There are always excuses for not reading: «I have a lot of work,» «the children are in exams,» «we are at the start of the production line.»

Napoleon took his books on campaigns, and these included history, philosophy, drama, and novels. Alexander the Great did the same, as did Winston Churchill. If they could, so can engineers and, more importantly, so should we.

Octavio Paz said: «When a society decays, it is language that is the first to become gangrenous,» adding: «As a result, social criticism begins with grammar and the re-establishing of meanings.»

That is why you must express yourself correctly, in an active voice and with the proper syntax. I do not expect you to know what these elements are, but be curious to find out and to be vigilant so as not to make unintentional mistakes due to lack of attention.

Unfortunately, no one will tell you that you cannot write or speak properly, they will simply select another candidate or give the project to someone else. It takes time, but by listening to speeches and reading copiously, you will realize the importance of expressing yourself properly.

Regardless of its usefulness in professional life, reading allows you to dream, and I think that is even more valuable. In 1875, Jules Verne wrote in *The Mysterious Island*: «I believe that water will one day be employed as a fuel, that hydrogen and oxygen which constitute it, used singly or together, will furnish an inexhaustible source of heat and light, of an intensity of which coal is not capable.»

For all of us who have read and reread Verne, his ability to captivate us with data and assumptions, and with prophecies, opened the door to dreaming of a better world than the one we currently live in (almost 150 years after Verne's book, fuel cells are beginning to materialize).

If you do not read, you will not learn or dream, and neither will you ever reach your full capacity. Science fiction is close to our world because so much of the equipment used in series such as Star Trek are now part of everyday life, e.g. voice assistants for devices (Alexa and Siri), electronic tablets to analyze data, the foldable phones that Motorola launched twenty years ago, and many other examples that are both obvious and still hidden in the pages of books. As an engineer, there is no excuse for not reading Verne, H. G. Wells, or Lovecraft, to mention just a few of the most emblematic. Start today.

Person
profes
develo

al and sional pment

A process that helps us to be better people in both our personal and working lives; that helps us to modify our behaviors and attitudes, make better decisions and overcome our fears; that pushes us to explore beyond our comfort zone and to (re)build ourselves as both people and as workers.

There is nothing impossible to him who will try.

Alexander the Great

20

IMPROVE YOUR PERSONAL TOUCH

Undoubtedly, engineers have a reputation for not being the most easy-going of people. We are often aptly described as withdrawn, uncommunicative, tactless, and focused only on our own business.

Of course, we are engineers, but that does not mean we cannot improve and leave the stereotype behind. To do that, we should try to loosen up, but that does not happen overnight. It is a matter of setting small goals and sticking to them; for example, talking to someone from another department every week and learning something from them too, however small it may be. It does not have to be a long conversation; the important thing is that you take the first step.

Certainly, you have to read up on the subject and seek advice: people will always be willing to support us if we ask for it in the right way and, vitally, if we state the reason. To convert to this type of behavior, the first step is to recognize that you have a problem. If you do not start from there, no change will be possible. And yes, I know what you are probably thinking: «I don't need it.» Big mistake. Maybe you need it more than anyone else.

21

LOOK FOR A MENTOR

Rarely in life are we recommended a mentor. Worse still, we think that finding one is something that «happens» without us actively seeking it out.

Who is a mentor? Someone who genuinely cares about your development and contributes to making you successful. It is a person who wants you to face challenges, to continually raise your level of challenge. In short, someone who brings out the best in you and helps you discover qualities you may not have even known about yourself.

Without a doubt, it is a luxury to be able to count on a mentor, but since they do not just appear, you must actively seek them out. Thinking of someone who can teach us is easy. We immediately have that intuition about people's capacity. This is necessary and must be coupled with an empathy for people and an awareness that part of personal success is luck. That is to say, we should not believe that things are obtained through effort alone: there are people who help, yes, but it also adds up to being in the right place at the right time.

I suggest that you have more than one mentor option, as the more successful they are, the more complicated their agenda, so you should have plan B, C and even D.

The hard work is finding that person, and then comes the easy part: asking for their support. However, arriving and saying, «I want you to be my mentor,» might immediately put them on the defensive. The better thing to do would be to ask for a recommendation that entails some subsequent effort on your part: «I think your expertise in strength of materials is outstanding. What books would you recommend on this subject?»

Then, read those books and send a note with your comments and, of course, ask for other suggestions. Alternatively, once the first piece of advice has been given, you can ask about a specific concern: «I want to study for a master's degree, what do you recommend?» or «I think I need to improve my written expression, do you have any ideas that can help me, please?»

In this way, the relationship will begin to strengthen naturally: there is no better compliment than being asked for advice. Again, seeking a mentor is basic for accelerated growth on a better path.

22

NEVER STOP BEING CURIOUS

José Saramago once wrote: «Old age begins when curiosity is lost.» Paradoxical though it may seem, today's young people are less curious. Perhaps this is due to the immediacy of the internet or the idea that all topics have already been covered.

The truth is that when we stop seeking new knowledge, an initiative that comes from a deep curiosity, we begin to age, and, given that the brain is a muscle, if not used, it atrophies.

Always ask yourself the «why» of things, no matter how trivial they may seem. Santiago, my son, likes soccer, and his team is Barcelona. I once asked him why they call their fans «culés», but he did not know the answer. Of course, he looked it up on the spot, and the answer popped up immediately. (Sorry, I do not plan on sharing it).

I then got curious and asked the same question to some friends who are also fans of the team. None of them could give me an answer. The world keeps turning, and this is just a banal example.

Nevertheless, the problem is aggravated when this type of situation is carried over into your professional life, and you lack the curiosity to learn or understand things. Reflecting on your lack of curiosity may help you find it again. Whatever the method, do not let the source of learning and wisdom dry up.

23

THE POWER OF «NO»

It may be the most powerful word, and perhaps that is why it is so hard for us to use: «no.»

When we answer negatively to something, our soul seems to immediately shrink, so we avoid it as much as possible. However, a «no» is able to free us from obligations that merely take up our time (the most precious commodity) but do not add value.

Not infrequently we accept future commitments because we see them as a long way off. We do not feel that they will have an impact on other areas of our life and yet they often cause significant losses at the end of the day.

Something as simple as an invitation to give a one-hour, seemingly innocuous talk can turn into two full days of preparation, travel, and answering questions from participants after the engagement is over.

Learn to say «no.» It may be complicated, but the first step is not to say «yes.» Shortly afterwards, you can answer that it will not be possible for a particular reason. As in everything, telling the truth works very well: «I can't accept, since I have previous commitments that I want to fulfill without affecting other priorities, such as spending the weekend with my family.»

Work on learning to say «no» and the horizon of your available time will soon expand. If you need encouragement, think of the many times you have been told yes, but then the commitment has not been honored.

24

CONTROL YOUR EGO

Engineering is not associated with ego as happens in other professions, such as being an artist, a soccer player, or a second level politician. However, given that so little is known about it, ego can be an evil that causes personal confrontations and bad decisions.

Since we study exact sciences in our profession, we tend to think that there is only one answer, with no alternatives; but a good engineer knows that there is always another way of doing things.

In fact, a common joke among colleagues is to ask what the difference is between a terrorist and a protection engineer. The answer is that you can negotiate with a terrorist. Like all jokes among colleagues, there is a lot of truth in this.

Another even darker facet of ego is when knowledge is not shared.

I am not referring to intellectual property or confidential information, but to data, formulas, and applications that we have gained through experience. Many colleagues take a position of not sharing, and they have every right to do so, but it is a folly.

Always think about giving back what society has given you. Do not let your ego dominate and be more powerful than you.

25

DEDICATE TIME TO THINKING

We engineers have a reputation for being square and inflexible, but with a superior mental capacity. We must prove things with facts, and this can only be achieved by thinking. Spending time on problems, without distractions, is the only way to find a correct, and even elegant, solution.

Achieving this is one of the greatest satisfactions. It takes discipline; it may take time to arrive at the solution, but you must lay the groundwork.

The answer may surprise you while you are swimming or doing another sport, that is how the mind works. Spend ten minutes a day thinking. You do not need more. Then you will see that ideas soon flow with greater speed and, more importantly, with better quality.

26

EXERCISE DAILY

The popular Greek phrase, «Healthy mind in a healthy body,» has been forgotten by a large portion of the population, and especially by us engineers. There are multiple reasons for this, among which are that our work sometimes takes us to isolated places, with little offer of healthy food or places to exercise, not to mention that wearing an industrial uniform makes changing sizes both simple and inexpensive.

But sooner or later, just like compound interest, the lack of exercise and neglect of our bodies takes its toll: the longer we take to pay, the more the debt exponentially increases.

Regardless of health, the other benefit of exercising is that it clears the mind, and this provides energy to carry out other activities. In addition, it allows us to sleep better.

Like any habit, it takes thirty days for exercise to mesh into our routine, but once established, the body demands its daily session. This does not have to be strenuous; walking is more than enough and an excellent start.

As with everything in life, there will be no lack of excuses, but avoid falling into these. Remember that if you do not invest some time in exercising now, you will pay for it later. Everyone makes the decision that they consider to be most pertinent.

27

THE COURAGE TO ASK

From a young age we were taught to keep quiet and not question our elders. Of course, prudence has its advantages, although it does not help when wanting to learn more.

There are no silly questions, but often the questioner is immediately criticized or harassed. Whatever the price, it is better than ignorance.

It is also true that sometimes we are reluctant to ask because we are afraid of being rejected. To solve this, you should just remember that you have nothing to lose. That is, it cannot be worse than your current situation. And if that were not enough, remember that questioning can also help to avoid risky situations. How many accidents happen because something is assumed rather than questioned?

Homework: the next time you are not sure about something, remember that there is not harm in asking, and that that is better than continuing in silence. There is a saying that argues that: «No flies can get in with your mouth shut.» This may be true, but neither can they get out. So, lose the fear of asking and practice it avidly.

28

YOU HAVE NOTHING TO LOSE

Never allow your insecurity to stop you from asking for what you need. It is impossible to get something if you do not ask for it and furthermore, the opportunity will not be there forever.

There will always be plenty of excuses for not asking; the important thing is that you take the initiative, otherwise, it is unlikely that someone will give you what you need.

Every day, we see people who receive more, and we attribute it to their luck, to the fact that their parents have money, or many other excuses. While these may be true, at least those people made the effort to ask. Every time you need to ask something or an opportunity presents itself, remember, «You have nothing to lose.» Do not hesitate.

29

LEARN FROM THE EXPERIENCED

Good engineers always seek to be close to their colleagues; this allows them to learn from the mistakes of others.

There is a saying that «no one experiments in someone else's head.» It seems to me that this is just a way to justify mistakes that are made daily, and a way to avoid facing reality. I think that «only fools do not experiment in other people's heads» reflects the idea better.

This, of course, includes reading and understanding the mistakes and successes that have been made. When we learn that Thomas Alva Edison made more than 1,700 different filaments before he had a functioning light bulb, we realize that, in addition to his intelligence, he had an indispensable quality: perseverance. He knew: «Genius is 1% inspiration, 99% perspiration.»

In addition, considering the cost of mistakes, enriching yourself from the experience of others is the most economical way to become a better engineer.

30

ON HIGHER EDUCATION

We generally study to obtain a higher education degree and to have a means to support ourselves; but it is also important to prepare ourselves to live. If you consider your profession only as a way of satisfying your economic needs, I can almost certainly assure you that you will be mediocre as an engineer and even worse, unhappy.

A large part of the good fortune of having a professional degree is that we form part of the top 1% of the population (discounting degrees from low level institutions).

That is why you should not see it only in utilitarian terms: what a waste! It is essential that you take actions that go beyond the simple performance of a profession, to have a more intellectual objective that allows you to obtain greater satisfaction from what should never become an obligation (work). A final piece of advice, go back to your *alma mater*, visit the place where you studied and remember how much you owe it.

31

RESILIENCE

A word that is currently in fashion but that only entered the Royal Spanish Academy (RAE) dictionary in 2014: resilience.

Philological history aside, what it conveys is the ability to recover from our defeats (which we have all experienced). If an engineer has not had any, it is certainly because they have not tried anything out of the ordinary.

Being resilient implies, above all, humility. Without this great characteristic we will always have some excuse that keeps us away from the solution. Humility gives us the strength to face the challenges we encounter and, of course, to look for new alternatives.

As an engineer you must face all challenges with humility and never confuse this quality with surrendering or keeping your head low; these are completely different things.

Obstacles are those frightful things you see when you take your eyes off your goal.

Henry Ford

32

LEARN TO LAUGH AT YOURSELF

Having the ability to laugh at ourselves makes life more bearable and allows us to face problems with a better attitude, in addition to facilitating interpersonal relationships.

We engineers tend to avoid it, and to avoid situations that put us in the spotlight. That is why it is hard for us to laugh at ourselves, it is in our DNA to believe that «we are infallible.»

What should you do? The first step is to recognize and know that your defects are also strengths. Are we not very sociable and talkative? The counterpart is that we may be reliable and discreet. Do we give accurate figures to the seventh decimal place? We are precise.

Being able to joke about our profession makes us more human and enables us to get more out of others. Here are a couple of witticisms to start this process: «What does an engineer use as a contraceptive? His personality.» «Engineer: someone who gets excited about things that others don't care about.»

Of course, there are hundreds of jokes, but these examples allow us to understand a little about how others perceive us. And do not forget perception is reality.

33

COMMUNICATE YOUR WEAKNESSES

A true engineer not only knows their weaknesses, which is already an important step, but communicates them without half-truths and with absolute clarity.

This attitude, which many may interpret as weakness, is actually the opposite, and even gives certainty to customers. Let me give an example on a personal level. The preparation of Grid Code studies, to which we devote a large part of our activities, comprises three areas: Power Quality, Short Circuit and Protection Coordination, and Information and Communication Technologies (ICT).

To the client, I unequivocally acknowledge that my area of expertise and my work for decades have been, and always will be, Power Quality. I also add that I know the subject of protections and short circuit quite well, and with the required level. I conclude by saying that I rely on experts for ICT, mentioning the experience they have.

Customers appreciate transparency, and they also understand that quality engineers are reflected in an appropriate price.

Being clear and direct in your communication will allow you to gain trust and avoid misunderstandings. Do it without fear, recognizing your circle of competence.

34

IF A THEN B

An engineer must see the consequences of their decisions, far beyond the obvious or immediate. If A then B is straightforward. It does not represent a major intellectual challenge. However, if A then B, then C or D, or maybe E, F, G, and H adds another level of analysis and professionalism.

Staying at B reflects a limited mindset and little intellectual rigor. For example, in the design of an industrial plant, we can consider that we are fully complying with the standards; that is, I design A and I comply with B.

Reality, unfortunately, is more stubborn and we know there are other points to consider. Will the load increase? Does the region in which we are installed have growth that affects short circuit values? Are the personnel who operate the equipment sufficiently trained? Will the contractor install what we have designed, or do we know that they tend to take a certain «license»? Will the piping withstand the pressure? Does the oil have the required parameters?

Of course, if you analyze *ad infinitum* you will never do anything, but that is no excuse for not acting and doing a thorough and rigorous analysis. You should always ask yourself: what if? what then?

35

BELIEVE IN YOURSELF

There is no heavier burden than the one you impose on yourself. If you think you can, you will be right, and if you think you cannot, you will also be right.

As engineers, we should always have a positive mindset and believe that the goals we set for ourselves are achievable. Unfortunately, many colleagues and peers do not develop even close to their potential because they mentally block themselves and put up barriers that do not exist. And if we add to this the false Mexican modesty, what we have is servility.

With decades of experience I can affirm that Mexican engineers have nothing to be envious of regarding our foreign peers. The huge difference is that they believe in themselves. «I can write a book», «I can solve this problem» are a couple of examples of the ambitions that we have and want to achieve. Do not be afraid to take the step (or steps) towards the goal you have set for yourself.

36

LEISURE TIME AND IDLENESS

We usually associate idleness with doing nothing or consuming our time in superfluous activities. Of course these are part of it, but I consider that not taking advantage of time is, perhaps, the main component of idleness.

Playing with our smartphone while standing in line somewhere is a common example of wasting time instead of using it to learn something. The same happens when we travel by plane: there are better things to do than watch a movie because, unless it is a long trip, we are unlikely to even finish watching it.

As engineers, we know the effect that the accumulation of an action has. For example, a 10% over-voltage decreases the life of equipment by 50% - it is not a linear result. When you take advantage of the time slots you have and learn something, the effects are not linear, but exponential. Do not miss the opportunity to maximize your preparation.

Seeking recreation is recommended and even mandatory. Disconnect, refresh the mind and, together with routine exercise, balance the workload. However, just because you have some «time off» from daily work does not mean that you cannot, and should not, take advantage of it.

For this, try not succumb to the first activity that comes your way. It is simpler to fall into the commonplace - watching a meaningless soccer game, for example - and do what the vast majority do. Do not forget that you can choose to keep watching a game or learn new words, perhaps perfect a language. As an engineer, it would be good to know the history of the profession (it is almost a must); this knowledge also gives certainty to clients and colleagues that they are dealing with someone who is truly professional.

Relating to the anecdotes of the «War of the Currents» is an intelligent and good conversation. You can talk about how Edison and Tesla clashed, the first defending direct current (DC) and the second defending alternating current (AC). Edison went so far as to electrocute animals to demonstrate the danger of AC (his biggest piece was the elephant Topsy, at the Chicago World's Fair). There are hundreds of examples, so as far as possible, find a hobby that contributes to your development but also helps you to disconnect.

37

METANOIA

This word, not very common in our environment, comes from Greek, and among its different meanings is that of «change of opinion or perspective.»

For many it may be difficult to change position, but it is essential to have an open mind to doing so. As engineers, this is even more important because the lives or welfare of others depend on our decisions.

Some years ago, I informed a customer that the switch in a harmonic filter would fail. However, the German switchgear supplier stood their ground. Reality can be cruel, and as expected, a failure occurred with a major explosion. Fortunately, the customer had taken precautions to ensure that personnel were not in the vicinity, which prevented a tragedy, and left just material damage.

Sleep on the decisions you want to make and do not respond to anything by thinking that only your position is correct.

38

«NEVER WASTE A GOOD CRISIS»

Much of this work was written during the COVID-19 pandemic. When lockdown began, few of us thought it would last so many months, but I remembered a phrase by Winston Churchill: «Never waste a good crisis» and I set myself several goals.

This attitude not only allowed me to make the lockdown more bearable, but it practically went unnoticed. Besides writing this book, my goals focused on having more time for myself, delegating, and giving up some activities.

During a crisis, most people tend to become paralyzed. Do not join the herd.

39

ON HOW TO BELIEVE

Jules Henri Poincaré, one of the most extraordinary scientists who ever lived, rightly said: «To doubt everything or to believe everything are two equally convenient solutions; both dispense with the necessity of reflection.»

Besides having discovered the theory of relativity before Albert Einstein, although he did not publish it, his main merit was his humility and his great common sense.

Think and do not just accumulate knowledge because, as Monsieur Poincaré also points out: «Science is built up of facts, as a house is with stones; but an accumulation of facts is no more a science than a heap of stones is a house.» Warehousing data is a misguided and very common behavior in most engineers. In this regard, let us remember that Montaigne used to say: «Nothing is so firmly believed as what we least know.»

Therefore, before starting any argument, understand the other party's level of knowledge. If it is low, try to keep a cool head and, as far as possible, avoid interaction. I have experienced this firsthand on the subject of the Mexican Energy Reform: everyone is an expert.

If keeping calm is not possible, set the rules beforehand and stick to them strictly. Remember that it is more important to be happy than to be right.

40

TRAVEL, TRAVEL, TRAVEL

There is nothing that broadens horizons as much as travelling to other countries and cultures, together with the interaction we have with the inhabitants of the places we choose to visit.

Go with an open mind, take risks (never use tours), talk to whoever you can, read about the culture of the place and, as an engineer, visit associations and organizations related to our profession.

At the end of this book, you will find a section with a brief list of museums around the world that every engineer should visit. Do not go for the safe and conventional in terms of travel; investigate and be surprised.

In Paris, for example, the Musée des Arts et Métiers, is a true marvel. There you can see Lavoisier's laboratory and Leyden's bottles, the forerunner of the capacitor or electric capacitor. Is it superior to the Louvre? No, of course not, but I guarantee you will leave hugely satisfied by the visit.

In Florence the Galileo Museum, next to the Uffizi Gallery, is well worth a visit. There are many more examples, see the list of those I suggest. Remember nationalism is cured by traveling.

41

THE PROBLEM WITH GOOD INTENTIONS

It is part of human nature to value well-intentioned people more than those who intend to harm us. However, it is not clear which causes more harm: Good intentions + stupidity or bad intentions + intelligence.

As engineers, we should always overlook (or at least minimize) the effect of motives and focus on facts. In Mexico, we have suffered considerably more from well-intentioned but foolish people than from ill-intentioned ones.

Many of the decisions that affect our lack of infrastructure, for example, were taken by individuals who «thought» it was the best thing to do, without having really analyzed it. Avoid falling into that trap. Blaming everything on corruption is an easy way out, but it is not the root cause.

42

EVERYTHING CHANGES (MORE THAN YOU IMAGINE)

Many of the mistakes we make do not come from not knowing, but rather, from not wanting to know.

Not wanting to know is not the result of ill will or selfishness, but of letting our thinking focus only on the problem in front of us. And this strategy does not generally come to a solution.

How can we avoid it? When approaching a problem, always keep in mind what it is that you are wanting to solve. It sounds simple, but it is not. By changing one variable you will influence others that are also part of the conundrum.

For example, generally in the business world, costs and benefits are on different sides of the «field,» to use a sporting term. If you want the design to be more robust, the cost could put you out of the market; but the opposite, a very inexpensive product, will not perform well.

Here's a word of caution. We engineers tend to give in too quickly to pressure from financiers, which may be valid, but the results could be disastrous.

I remember early in my career, installing a Digital Equipment Corporation computer system, using ten-inch floppy disks that were extremely dust-sensitive. Roof-top air conditioning was not authorized, and a window unit was installed. It failed continuously, and since it was a non-stop process for the manufacture of glass containers, calls in the early morning hours were constant.

When the second machine arrived, I once again requested what we required, and the same result was obtained. The solution was found by calling the buyer's house to inform him that we had had another stoppage. On the third call, as you can imagine, aircon was authorized; of course, I took the opportunity to buy two and to be able to sleep peacefully.

Moral of the story: wrong decisions have to be felt by the person who made them. The sad thing is that each stoppage cost more than the equipment and no one in management took the appropriate measures. That is when I knew that my stay in that company would be short.

43

CANNONBALL OR ROCKET?

In life you have the choice of behaving like a rocket or a cannonball. In the latter, once fired, the trajectory depends entirely on the laws of physics.

A rocket is different as it is controlled either remotely or by a pilot, allowing adjustments to the flight plan if it deviates from the original path.

Unfortunately, most decisions are made cannonball style. We might think this is strange, but the root cause is that by not looking at the consequences of our actions, we maintain the illusion that we are capable and never review the effects of the decisions we make, perhaps thinking that the job is complete. We turn around and engage in new problems and while the burden of responsibility may have been lifted from us, we have not solved the problem.

44

YOUR DETERMINATION DEFINES THE RESULT

Nothing is as important as having the determination to achieve something. There are no alternatives if you really want to achieve your goals in life. It is a quality that will always make a difference. Nevertheless, the true reflection of not having determination can be seen in some of our athletes: they diminish in the face of challenges.

If there is determination, barriers fall like dominoes, one after another. If you do not give up, you will surely succeed. Will and perseverance deliver.

45

THE MORAL RESPONSIBILITY FROM STUDYING IN A PUBLIC UNIVERSITY

Most engineers in Mexico have studied at universities or institutes that operate with public funds.

This means that they must return to society what they received, practically for free. To ask for more opportunities, benefits or handouts is to have some nerve.

Unfortunately, the system makes many students consider themselves victims when in fact they are quite the opposite. I have taught in all types of institutions and, sadly, apathy is the same in both private and public schools. The huge difference is that in private schools, the lack of achievement is a matter between the parents and the student. On the other side, it is between society and the students.

Do not get confused and accept your responsibility.

46

PREPARE YOURSELF FOR THE FUTURE, NOT FOR THE PAST

Most of the lessons we receive are focused on the past. When we leave our studies and face professional life, the reality is completely different. This applies not only to receiving a lesson, but also when we give one. Keep this in mind. Conduct yourself as you would drive a car: looking straight ahead and not in the rearview mirror.

47

KNOW WHEN TO WALK AWAY FROM A PROBLEM

More often than we may want to recognize, the professional pride of engineers provokes a stubbornness to solve a problem that may not have a solution, or at least, not one we can see in the moment.

Knowing how to abandon a problem is not a lack of discipline, it is intelligence. If the decision has been deeply analyzed, it also does not imply a lack of tenacity.

48

WHAT CAN YOU CONTROL?

The author of *Gulliver's Travels*, Jonathan Swift, was also a great essayist; his most remembered phrase is perhaps the following: «When a true genius appears in the world, you may know him by this sign, that the dunces are all in confederacy against him.»

The problem with trying to control what others do or think is that you waste the energy that you could be using to focus on the only thing that you can really control: your decisions, your attitudes, and your actions.

Fighting something you have no control over is a waste of energy and skills that should be directed toward productive activities. If you think this does not happen to great men, remember Edison and Tesla, who fought each other unsuccessfully.

Analyze whether your decisions have been taken calmly and unclouded by envy and lack of understanding. It happens very often.

49

THE IMPORTANCE OF THINKING

Among many other contributions, René Descartes bequeathed us the coordinates that later acquired his name: Cartesian. As a good man of the Enlightenment, his knowledge was diverse.

His philosophy was synthesized in the phrase: «I think, therefore I am,» which largely sums up his totally rational view of the world. However, I prefer Nicole Oresme's phrase: «I doubted to be, therefore I am.»

I think that doubt is what leads us to analyze the different options and routes we can take. When I speak of doubt, I am not referring to inaction, but to questioning what we think and what we do. This is basic to reflecting before acting. Thinking is good, but we can be wrong.

If you have the discipline to reflect (think), even briefly, before acting, you will have excellent results and mistakes will be reduced.

By the way, Nicole Oresme lived 300 years before Descartes.

50

ABOUT DECISIONS

«You can't make an omelette without breaking eggs.» This idiom confronts us with the inevitability of decisions, as does the Mexican version: «You can't whistle and swallow pinole.»

Knowing how to decide is one of the most important tools for prospering. Remaining static and thinking down to the smallest detail will never make things happen. This is called paralysis by analysis.

Generally, when facing an interesting project, our enthusiasm drives us to look for options and good reasons to do it. However, the time will come when you have to do without something to achieve the final goal. The key is to understand when you have to act and then do what is necessary.

In the business world there is a phrase that I like very much: «Strategy that doesn't hurt, is not strategy»; that is, if you don't make tough decisions - leaving something you're craving, for example - it won't really be a strategy, just wishful thinking. Make no mistake.

51

«YOU WILL NOT BEAT THE EAGLE»

Although I am an inveterate reader of phrases and proverbs that condense popular culture, I was surprised when engineer Rogelio Gasca Neri, former CFE General Director, told me the one that I allude to in this tip (the Mexican flag has an eagle in the center).

We were talking about the confrontation between private individuals and the government, as well as the imperative to negotiate. Unfortunately, when tempers flare and lawyers stoke the fire, the message is that beating the Government is extremely difficult and if it is achieved at all, the victory will be pyrrhic.

Therefore, not only in negotiation with the Government, but in any negotiation where there is an asymmetry of power, it is better to seek an arrangement that suits both parties. When I was a student, we all knew that confronting a professor was not a good idea: sooner or later you would move on to another subject or one of their colleagues would even the score.

I am not talking about negotiating with cowardice, but rather, with intelligence and measuring the consequences. A good engineer must not only know how to calculate circuits, but much more.

52

GET CLOSE TO POLLINATORS

Our natural tendency is to approach the majority to learn from them, but we forget that the minorities are the ones who can teach us the most.

Therefore, a good engineer seeks to relate to minorities that bring different things and see reality from a different point of view.

What activity has few followers? Chances are you will find more capable and interesting people there. Try to be unconventional. It pays off handsomely.

A typical example can be in university. Try to talk to students from degrees that are not so popular, as you will surely learn a lot from them.

What can I learn from an accountant? From a doctor? From an architect? If you don't question yourself and put it into practice, you may miss the opportunity to get another perspective that will probably surprise you.

53

HAPPINESS IS NOT THE GOAL

As Albert Camus wrote: «The struggle itself toward the heights is enough to fill a man's heart.»

In other words: we will never achieve total happiness; it is not possible. However, what we can do is enjoy the journey in trying to achieve it.

If you live thinking that you have not reached it, you will be irremediably unhappy. That is why an engineer must set appropriate goals and progress towards them. For example, learning about fire protection, or fluid mechanics, or resistance of materials, even if they are not our main activity, will allow for small triumphs that, added together, are the secret to happiness.

Do not despair, just try to make daily progress, however small each step may be. If someone asks you if you can run a marathon, you will most likely say no; but if they ask you to do 42.1 kilometers divided into 500 meters per day, the answer may be yes.

54

HOW ARE WE TO BELIEVE?

If anything has characterized our society in recent years, it is polarization and the defense of ideas, without questioning whether they are true, half true, half false, or false. They are assumed as absolute truths and that is it, making dialogue impossible and disqualification immediate.

For questions of politics, vaccines or whatever topic you choose, people stand firm in their beliefs, do not listen and therefore are convinced they are right.

How to decide what to believe in? We will call the first method tenacity. Choose what to believe and stick to the idea regardless of logic or evidence. The next is that of authority, believe whatever the political, religious or other authority says about what you should accept at face value.

The third we will call an *a priori* approach, and it is about being reasonable and believing what sounds good to you, even though it is not necessarily based on facts. You often end up believing what the majority believes, even though it may be completely wrong. The fourth, typical of most engineers, is the scientific method. This consists of accepting the fact that you can be wrong, and, given this possibility, reviewing your basis and accepting if you are not right.

Unfortunately, the first three methods are the most often used and, moreover, are very easy to apply; not to mention the feeling of satisfaction they create for «being on the right side.» Learn to doubt and always seek the truth.

55

NO REGRETS

Living by the philosophy of «I have no regrets» may seem attractive as it is impossible to change the past, and is why many authors and counselors advocate that we should not waste time on regrets.

In fact, regretting something is a natural and healthy feeling as long as it is not abused. Well used, it can be very valuable: it helps to clarify things, instructs us and, well managed, can lift us up. For example, instead of regretting not having studied a subject in depth, you can look for someone to support and guide you to master it.

It is also helpful to discuss your regrets with someone and receive support. Contrary to what we may think, when we express our weaknesses or past mistakes, people tend to respond in solidarity and with healthy advice.

Perhaps the most useful thing to do is to learn a lesson from it. If you try doing this by focusing on yourself, it is likely to be difficult, so an effective method is to place a friend in the situation: imagine that they are the one facing the problem and take the learning from there.

An example in engineering can be when you make a mistake, let's say you miscalculate the gauge of a conductor. If you place a third person in this situation, rather than yourself, you will analyze the situation more clearly and learn useful lessons.

It is also good to have a list of people who can support you in these cases. This way, when the need arises, it will be easier to ask for help. In a crisis moment, it is more difficult to find help.

56

THREE RULES FOR EVALUATING IF INFORMATION IS TRUE

In this hyperconnected world, it is easy to believe the first thing that appears on social media or in a search engine. It is comfortable because we do not have to think much, and we avoid questioning.

To change this attitude, the first step I recommend is to check your feelings: Do I believe this? Does it seem reasonable to me? Learning to listen to what your heart is saying can help you clarify. Second, ask yourself: what is my personal experience? That is, have I ever seen or heard something similar from a reliable source?

Finally, you have to give yourself time to ponder the information and not take a position prematurely, give yourself space from the issue you are analyzing and come back to it later. Remember Umberto Eco: «The truly genuine problem thus does not consist of proving something false, but in proving that the authentic object is authentic.» This takes more effort, of course, but it will avoid a lot of headaches.

57

FACE THE EVIDENCE

It is understandable that when faced with data or facts that we do not like or that affect our interests, our position is to deny or minimize them. The opposite is also true: when the evidence validates us, we place greater emphasis on it and avoid questioning it.

This is an «ostrich effect,» i.e., we hide our heads so as not to have to face unflattering news or results that go against our interests.

One technique to lessen this effect might be to ask yourself, before analyzing the information, how it makes you feel.

If it causes you uneasiness or discomfort, it is a sign that you should be careful in the process of reaching final judgements. This decreases the possibility of confronting the data with the (conscious or unconscious) desire to reach a conclusion that we like.

What follows from this is that we are more likely to see what we want to see. Let us remember Molière, French playwright and great creator of satires: «A learned fool is more a fool than an ignorant fool.»

**A man
is but a product
of his thoughts.
What he thinks,
he becomes.**

Mahatma Gandhi

58

HOW TO OBTAIN THE ATTENTION OF THE PUBLIC

Communicating in engineering is complicated, to put it mildly. Three-phase system? Entropy? Molecular equilibrium? Where do the electrons go? Real and apparent power? For those of us in the sector, the above concepts are simple, but for those who have not been involved in this area, the opposite is true.

So how do you attract an audience to unfamiliar topics? The first thing is to stimulate their curiosity and that can be done by asking questions. When I explain the analogy of an electrical power system, I do not start with: «Next, I will explain...,» but with questions, like «do you know what is the greatest machine ever created by man?» No one, so far, has answered correctly. By the time they hear the answer I have the audience where I want them: eager to learn.

But it does not end there. We have to continue to pique their curiosity. On the subject of frequency, when I ask how many electrical cycles there are in a year, many want to take out a calculator, but I ask them to rather think about it. When I mention that the answer is greater than 1,800 million, I can see the surprise on their faces. In short, you must put yourself in your listeners' place and think hard about what would catch your attention; that way you will know how to get the audience in the palm of your hand.

59

ON THE DIVERSITY OF OPINIONS

It is enough to go to any bookstore or newspaper and magazine stand to see the plurality of opinions we have in Mexico. In Paris, for example, I found five newspapers. Here there are easily fifteen, and no doubt there are more. The same goes for magazines and books.

This is logical. We are, possibly together with India, the most pluralistic society. We have at least sixty languages, without taking such absurd positions as in Spain where they think that a language makes a country.

This leads us to the fact that in any matter, both personal and professional, we will have a plethora of different points of view and ways of looking at reality.

In the north of the country, they do not understand passivity in decision making, just as on the other side they do not understand the ambition to emulate the «gringos» at all costs.

What does this have to do with engineering? Well, you have to adapt your instructions and expectations to the place in which you find yourself. It is vital to not expect the same response, for example, in one place as in another. Trying to change people is a waste of time. If anything, try to make them a little more receptive to what you want.

Foreigners complain that Mexicans lie. They do not realize that they are being told what they want to hear; if they believe it, it is their problem, not ours.

Think about the environment or area you are in and adjust. Do not waste your time trying to change reality.

60

CREATE ANALOGIES

Few educational tools are as powerful as a good analogy. Its goal is to enable a twelve-year-old child of average intelligence to understand a complex subject without a problem. It sounds simple, but it is devilishly complicated. It requires hours of thinking, analyzing, and listening to subject matter experts to improve the model.

Creating and refining the analogy of an electrical power system, the largest machine ever created by far, took me more than three years, and there are still areas for improvement or where new ideas can be incorporated (see the following link: https://goo.su/qtQgv5n).

In addition to the satisfaction of having created this model, the feedback from the «sacred cows» of engineering was surprising. I quote: «This should be the first class in electrical engineering. People should know that we are here to make the bike run smoothly.» «I would recommend the Barcón bicycle.» «Thank you for simplifying something so complex.»

Undoubtedly, analogies will help you reinforce your knowledge and develop a broader vision of the subject. Be sure to look for them in your presentations, dialogues, and debates. You will have a huge advantage.

61

HOW TO CREATE THE FUTURE?

What tomorrow holds for us is beyond anyone's knowledge. We can make guesses, all equally ineffective, from hiring an expert to consulting the horoscope.

There are hundreds of methods, but the only sure way is to deliver everything in the present; besides, this is the only way to achieve recognition and a certain degree of protection for the future.

So, worry less about what is coming and focus your efforts on the present. This is even more important from the viewpoint of the young: what you do today will be part of your future, so do not look with disdain on the fact that sooner or later you will grow old. It may seem far away, but one day it will suddenly be right in front of you.

62

ON ATTITUDE

It is our attitude, not ability, that determines how far we will go. It is extremely enlightening to read about the experiences of the great engineers who came before us. Undoubtedly, their intelligence level was far above average, but the common denominator between them is their attitude.

How many geniuses have wasted their talent because they lacked a positive attitude? This characteristic helps you to persevere, and without that, you will not make much progress. Do not be intimidated by problems but instead, face the challenges ahead of you.

63

DO NOT CONFUSE WANT WITH NEED

We tend to use «I need» as a catch-all for our desires. Let's think very carefully about when it is a need and when it is just a craving. Otherwise, we will get nothing but denials and frustration.

You do not need a new cell phone, you «want» one. It is valid to desire, but you should not get confused because this will not contribute anything to your life and will only cause dissatisfaction.

64

PROPOSE AND DO NOT COMPLAIN

Complaining is perhaps the most supine waste of time. In reality, no one really cares about the problems you have, and, most of the time, when you raise them, it is to seek sympathy in the face of inaction. «You whine and complain but allow it» is a phrase that sums up this attitude very well.

An engineer who complains can never be successful. The reason is simple: when you complain, you limit your brain. Instead of using it to think about how to change, you use its processing power to find novel ways to complain or justify yourself.

The difference between a first-rate engineer and a mediocre one can be summarized by the fact that the former proposes alternatives to a problem and the latter only talks about the problem.

If you think something is wrong, you have two options: complain or propose a solution. The second should be the only alternative, but the first is the most common.

Do not complain about your boss, rather propose changes, do things differently and look for alternatives. If after a certain period of time everything remains the same, consider resigning. Complaining about a complicated procedure can only be fixed by simplifying it, and for that you have to make a proposal.

There are circumstances in which one cannot wait for change to occur gradually. Let's take a security aspect as an example. It is not a matter of proposing, but of acting and refusing to continue with the work. On several occasions, I have seen operators and their bosses in precarious conditions with no action taken, even though it is known that they are at serious risk.

In short, a good engineer does not complain, they act. Our profession works in real time, and that is how we should proceed.

65

ABOUT DIVERSITY

An engineer should constantly seek diversity. Only in this way will we expose ourselves to new ideas and build innovative thinking.

Although it sounds like a common refrain, the world is becoming more diverse every day. We see it in all sectors, and it relates not only to the proportion of women and men, but also to phenotypical characteristics. I remember an image of the German national soccer team. For decades, it was impossible to see a player who was not white. The same was true for France or Italy. Now we have a multicolored mosaic. The same thing is happening with respect to gender in universities, workplaces and in management positions. It is a trend that has no return.

The best way to get used to diversity is to experience it. Go to places you do not normally go, visit places far from your usual activities, study other subjects, and look for opportunities. Change your daily routine, walk down different streets, see movies from other countries, or go to a different bar. There are dozens of ways.

Of course, it is also useful to read about other cultures and ways of life. Videos help, but the degree of immersion and understanding we get from reading is completely different: do not take the easy way out. Eat new dishes, order a different beer, listen, even briefly, to a Mozart opera. You will be surprised by how your panorama widens.

You should read about different topics, attend events (even if they are not so motivating), listen to opposing voices with an open mind, attend exhibitions that are not in our field, enter stores that sell different and original things.

One of the best experiences I have had was attending Gift Expo. Over 90% of what was on display was, from my perspective, impossible to sell. To my surprise, customers thought differently and purchased what I rejected because I assumed it was junk.

I learned something from that: everything can be sold. This changed my perception of trying to tell a client what they need (a very common mistake in our profession), rather than giving them what they ask for.

On the other hand, without the diversity presented at modern art shows it would not have occurred to me to design capacitors in a two-color cabinet. Of course, getting the distributors (colleagues, of course) to accept it was a tough battle. We finally did it and the result was beyond what we could have imagined: customers bought it just because they could put it in their stores or on their outside walls without hiding it.

This leads me to emphasize the importance of design in our profession. From now on, whenever you consider a new product, think about the design from the start.

Profes
practi

sional ce

Activities, situations, and particular interests directly related to the work-professional task; with a specialized knowledge that also determines our entrance or belonging to a specific social group.

Anyone who renounces their vocation for 'practical reasons' commits the most impractical idiocy.

Mario Vargas Llosa

66

SAFETY

For an engineer, the first obligation is safety. Nothing and no one should stand in the way of this cardinal point. We start from the premise, which can be no other, that all accidents are avoidable.

There is no exception, there is no such thing as bad luck or providence. Every accident occurs due to a lack of precision and a failure to take into account each and every possibility, in particular, those related to irrational actions on the part of the operators, although these are not the only ones.

Several years ago, an employee died from electrocution in the company where he worked. He had jumped over a more than four-and-a-half-meter fence to get his jacket. Yes, he was the operator of the high voltage test laboratory, he knew all the risks, he was trained, and his behavior had been impeccable.

The local authorities (in this case the Social Security) and the insurance company ruled that the operator was solely responsible. The direct bosses were understandably distressed; no one likes to see a colleague die, let alone think that it could have been avoided.

A week after the event, I took them to the lab and we identified various security actions that, had they been implemented, could have prevented the accident. These included extending the fence up to the ceiling, a presence sensor with alarm, video cameras with recording, authorized access by security management, and even bigger signs. Six months later we went back to the lab and still came up with new ideas.

The lesson here is that safety has no finish line, but a road you travel with no end. Of course, after the accident, we not only reviewed the laboratory, but the entire plant. The process continued with defensive driving courses for employees, as well as advice on how to avoid accidents at home.

Another anecdote I remember is when I worked for a Swedish group. I was proud of the safety system we had in place, but after the inspection of the plant, the area director, who later became a mentor, commented that my evaluation would mention that we were not at the required level in terms of safety.

When I questioned him about this, his answer (which I will never forget) was the following: «When we left in your car after breakfast at the hotel, you started the car and only then put on your seat belt.» Of course, I replied that when that happened, I had not yet moved the car, but he only answered: «Safety first.»

To reiterate: all accidents, without exception, are preventable. And there is no greater responsibility than the well-being of people.

67

THE BOSSES

There is nothing that impacts our professional development more than a boss who allows us to grow, teaches us, and has the potential to become a mentor.

A good boss guides us, which is always important, but even more so at the beginning of our professional life.

Like good teachers, we may not always like our boss very much because they make us suffer, they may demand more of us than we think we are capable of, and they show us that we have much yet to learn. Naturally we hate them, but in the end, we are grateful for what they have gifted us.

It is easier to select a good teacher, we already know them and we may even have studied subjects with them. However, it is difficult to identify a good boss, but not impossible. You can learn a lot in the initial interview, and you also have the opportunity to ask questions; these should be careful and respectful.

Address their people management style, ask how many people report to them, how much time they spend developing their team, and whether they continue to develop themselves, among many other things.

Luck rewards those who prepare, and it pays to study the various books on how to conduct yourself in an interview.

Remember that work interactions are not only about providing information, but also about obtaining it. In this sense, if while at work you realize that the boss is a pain, it is time to quit.

Do not waste time or delude yourself that they will change and that things will get better. It will end up hurting you because you will not improve the boss, but they will prevent you from developing your potential.

We must take care of time; it is the most valuable thing we have, and it is the only thing that cannot be recovered.

68

ON STANDARDS

When you hear an engineer say, «The equipment meets the standard,» run away and do not turn around! The standard is the *mínimum minimorum*, a Latinism meaning the minimum necessary or the minimum of minimums. In other words, anything less than that is unacceptable.

Any self-respecting engineer knows that the standard is only a reference and that, from there, the analysis is based on what is required. For example, the temperature values set by the International Electrotechnical Commission (IEC) for capacitors with low voltage power factor correction are not adequate for most of our territory; however, they are offered and not questioned. Perhaps the bombastic name of the Commission intimidates engineers and that is why they are not able to question its validity in our reality.

Add to this the fact that Spanish-speaking engineers are not very good at writing, so it is easier for us to take refuge in an international standard than to write, adapt, and modify what is needed.

69

SOCIETIES AND INSTITUTES

Keeping up to date is not an easy task. Responsibilities that seem to increase daily, in many cases involving the family, make this goal difficult.

I stress again that, unless further study is done, sooner or later obsolescence will occur in the same way that humidity attacks iron: gradually, until it is permanently weakened, and then there is no way to return it to its original state.

For this reason, one of the best alternatives for reinforcement in this regard is to belong to professional associations, institutes, associations, societies or other such organizations.

The permanent contact with colleagues and receiving frequently updated material will make your work easier and more pleasurable. Not only do you receive, but with a minimum of reciprocity, you give back.

In the electrical field in Mexico the College of Mechanical and Electrical Engineers (CIME) and the Mexican Association of Mechanical and Electrical Engineers (AMIME), are both institutions with several decades of contributions to engineering in the country. Each has its strengths, and the choice depends on the area in which we are specialized.

In the international arena, but with a strong presence in Mexico, is the Institute of Electrical and Electronic Engineers (IEEE), the largest association in the world with more than 480,000 members. (Mexico is part of Section IX).

Their monthly magazine, *Spectrum*, is easy to read and covers the most diverse aspects of electricity and electronics. The articles are excellent as well, and membership in several societies is available. The closest to us are the Power and Energy Society and the Industry Application Society.

If your interest is more focused on high and extra high voltage, check the Conseil International des Grandes Réseaux Électrique (CIGRÉ); civil engineers have the Instituto Mexicano de Ingenieros Civiles (IMIC); chemists, the Instituto Mexicano de Ingenieros Químicos (IMIQ); mechanics, the American Society of Mechanical Engineers (ASME), among others. Each discipline has numerous options, and we should take advantage of them.

Cost cannot and should not be an excuse. We are talking about less than ten dollars per month. If this amount seems excessive to you, put your degree in a drawer and look for another profession.

70

LOVE YOUR PROFESSION

Few things are as despicable as those who do not love their profession, and I do not mean just a total and absolute respect for it, but something that goes beyond that.

You should know that doing things wrong tarnishes colleagues and that we all pay for those acts. You cannot proudly say «I'm an engineer» because you may not have done anything to earn it.

It is not a question of asking, but of acting. You must demonstrate it in your daily work. Leave no doubt that you will defend the profession to the hilt.

This, importantly, includes denouncing those who do not fully comply with their legal and ethical obligations and pointing out those who are on the borderline. Failure to do so makes you an accomplice and affects the entire guild.

71

TEAMWORK

To succeed, you have to be a team player. No matter how good we are as engineers, we will always - and I repeat always - need support from other colleagues.

Intuitively you know it, but that little demon called ego advises you not to try. Of course, you will have some setbacks and several betrayals along the way, but these are not reasons for you not to go through with it, the end result always outweighs the alternative.

Working as a team has many similarities to the amount of water we can hold in our hands: the more open they are, the more we can hold. A fist does not hold much water.

You cannot build a team (or work with it) if you always want to be in control and, even less so, if there is destructive criticism of the performance of other team members. Teamwork always involves using «we» and not «I.» It is also important to make sure that the group is diverse. Yes, I know, it is more comfortable to be with clones of yourself, but few creative ideas will emerge from a totally homogeneous group.

Perhaps the most important key is to ask yourself «who can do this better than I can?» Make a habit of always asking yourself this question and it is sure to lead to productive results.

72

ABOUT THE NEED TO REST

Every machine, no matter how precise or well built, needs rest periods. In fact, they are mandatory. The same goes for people, but I will focus on our profession.

We work in places where carelessness has serious consequences. Few remain well after an electrical accident and a considerable majority die. Therefore, we cannot be tired, unalert, drowsy or in any circumstance that prevents us from being at our best.

An accident is not always the responsibility of the fatigued person, nor does it occur immediately. A poorly wired board can fail months later, incorrect grounding systems will create potential differences. So, you cannot afford to not be in top shape.

Always being in a frenetic state of activity will not lead to good results. The brain needs rest, thousands of studies have proven it. Even if you think it does not apply to you, sooner or later you will discover that it does.

An important point: resting does not necessarily mean watching television or reading banal gossip magazines. These ways are valid, of course, everyone selects what works for them. What I mean is that if you dedicate yourself to writing or studying about a topic that you generally consider work, if you manage to disconnect this activity from the routine, perhaps it will become a rest.

As a corollary, the greatest discoveries did not come from being at a desk. They may have been cultivated there, but they appeared in the most unexpected way. Has it ever happened that you cannot remember a piece of information and the more you try to get the answer the less you can? Very simple, talk about another subject or do another activity: soon the answer will appear. This is how the brain works. Do not try to fight it, rather give it a rest.

73

USE YOUR HANDS

I believe that to be a complete engineer you have to use your hands.

There is no comparison between a desk job and action in the field in terms of the speed of learning. Moreover, the latter is highly pragmatic, since we are confronted with reality and not with a textbook.

This allows us to assess the difficulty of a task and, when designing a product, take that experience into account. For example, the space needed for comfortable wiring will be clear, and thus the design will not present problems in the future due to incorrect installation.

If you have never had to handle a conductor, the cabinet design will certainly decrease in size, as only electrical clearances will be considered; or as has already happened to me, some «engineer» thought my design was incorrect and overkill. Money could be saved. Of course, the market reaction was immediate: they returned the product as it was impossible to bend and connect the conductor.

Another advantage is knowing how to correctly assess the time required for each task. With the experience that you have gained through using your hands, you can decide that criteria. Do not let these opportunities pass by you.

Yes, practical activities are less comfortable and may carry greater risk, but the experience gained is worth it. Remember that we leave college with no practical experience, and the best school is life. I had the good sense to practice it and I am still thankful.

74

ATTENTION TO DETAIL

The Anglo-Saxons say: «The devil is in the details.» I agree more with St. Teresa of Avila: «God is in the details.» This phrase best describes the importance of paying attention to those things that may seem minimal, but that make the difference between a normal, acceptable job and a memorable one.

Engineering lends itself particularly well to careful and thorough work. Take a protection and control panel as an example. Something as simple as a neat, spaced, orderly, logical and well-marked wiring harness will functionally do the same job as one that is a tangle of conductors with no building structure.

The details may possibly cost more, but at the first maintenance or upgrade, the time and cost invested, coupled with the risk of an accident, will eliminate the savings. Another example is installation manuals. A well-written one, with proper syntax, without spelling mistakes and revised in depth, allows us to work with greater speed and safety. Videos certainly help, but it is essential to write a good script, otherwise it will be a real pain to watch.

Steve Jobs, co-founder of Apple, insisted that the computers' electronic tablets (which no one sees) should not only be functional, but also aesthetically pleasing.

This goes hand in hand with what José Ortega y Gasset said: «Clarity is the courtesy of the philosopher.» Paraphrasing him, precision is the courtesy of the engineer.

When a piece of equipment performs its function accurately, we thank those behind its design and manufacture. The same is true when the instruction manual has this quality.

I have always questioned why accuracy is so rare. I think the answer, to a large extent, lies in trying to do too much and not spending the time required on each thing. In other words, the work is done, but not polished: from draft through to production.

Of course, there are cost considerations and the strategy may be to deliver at the lowest price and let the client manage. My recommendation if you work in such a company is to quit. Do not waste your time and be clear that this will not allow you to grow as a first-level engineer.

«I am writing you a long letter, as I did not have time for a short one,» defines this circumstance. Since I did not take the time to polish my reply and make it concise, it is now your problem to understand what I mean. Examples abound, but the most important thing to remember is that precision in details becomes a habit. Not only at work, but also in everyday life.

75

RETIREMENT

The years fly by and if you do not plan for retirement, it is highly likely that old age will not be a great stage of life.

Popular culture says that bullfighters and artists do not retire, and, to a great extent, it is because they enjoy what they do. When work is enjoyable and enriching, it ceases to be work. Of course, day-to-day problems will remain, but when you have passion for what you do, small obstacles are easily dealt with.

I do not mean working until the day before our funeral or cremation, but rather, having activities that keep us active. Remember that if the brain is not used, it atrophies. You can teach classes, serve as a consultant, or sit on company boards, for example. There are hundreds of similar activities.

If the argument for staying active is not appealing, two additional points apply. First, staying at home is tantamount to breaking a pact. We are animals of habit and having someone at home when that person was not there before is a breeding ground for conflict. Second, something that few people consider: the retirement age has remained the same, but life expectancy is increasing every year. Do not make the miscalculation that I have seen by dozens of colleagues when they think: «I'm 58 years old; I'm retiring at 65 and life expectancy is 74 years for men.» What we do not quite understand is that figure includes those who died at birth and those who died by accident or heart attack. The fact is that at age 65, one can reasonably expect to live 20 to 25 years more in average health.

Pensions will not be enough to live comfortably, so we have to work for a few more years. As the advertisement of a Swiss private investment bank used to say: «Remember: the 60's of now is the 40's of before.»

76

«LET FOOLISH WORDS FALL ON DEAF EARS»

As electrical engineers, we know that many equipment performance problems are the result of improper grounding systems.

Therefore, a power quality study that does not include an analysis of this area is weak, to say the least. However, this is not a technical book, but rather, one of practical advice, and what I wish to communicate is that you must know how to «let foolish words fall on deaf ears.»

This implies that you should not get upset or annoyed by unwise comments, analyses, or disqualifications based on your gut reaction rather than your brain. If you are able to do this, your life will be better.

Over the past few decades I have seen such extreme blunders, largely encouraged by unbridled commercial interest, that nothing surprises me anymore: power factors of 1.35, magic earths (to continue with the theme), magnetic gasoline aligners, and dozens more.

So, «let foolish words fall on deaf ears» and answer without getting upset, without scoffing, and start from a deep technical analysis. This is the best recipe, and with it you will have won a large part of the argument.

77

MIRACLE PRODUCTS

My next book may be a compendium of the «miracle» products I have encountered. There are literally hundreds that promise extraordinary results that not even those advertised in the dietary arena - such as immediate weight loss to get the body of Adonis in just two weeks - can match.

Because of my line of business (electric power), the vast majority of these products come with unrealistic promises of reduced consumption and, as an added value, a long life in the system.

From the negative susceptor, Ecowise, and many others, energy savings of more than 40% are promised, immediately catching more than just a few unwary people. And that is just one example!

Of course, one can justify the buyer's mistake, since perhaps they do not know, but it seems to me a very paternalistic attitude. If something is so good, why is it not sold to big energy consumers? Why does it not have a patent? Why is its inventor not the richest person on the planet?

These questions do not imply any technical knowledge, they are common sense. Of course, the illusion of making easy money overshadows the exercise of asking these simple questions.

These are the questions I use to debate a product, rather than focusing on its technical feasibility, and the discussion ends quickly. Moral: not every technical debate is won with technical arguments. Think laterally.

Nothing is more difficult, and therefore more precious, than to be able to decide.

Napoleon Bonaparte

78

ENGINEERS IN CHARGE OF ENGINEERING FIRMS

We already agree that, as engineers, we have an obligation to defend our profession. This implies that engineering-related companies, such as the CFE, the National Institute of Electricity and Clean Energy (INEEL), the Mexican Petroleum Institute (IMP), PEMEX, as well as technological universities, technical schools and the like, should be managed by engineers as they will otherwise begin to become distorted and lose focus. There are several typical cases of this, for example what occurred with the CFE and therefore with the National Electric System, not to mention PEMEX.

In the case of the CFE, decisions were made by financiers or politicians, with the lowest net present value requested as an evaluation parameter. This resulted in a focus on generation plants, places with easy access to gas, and low altitude locations. The correct decision would have been to carry out adequate distribution across the national territory, and thus the system would have been meshed and a robust gas pipeline network would exist.

This bad practice continued with renewable energy installations, which the system does not support, but this does not matter because it is extremely profitable for those who own it. In short, decisions of such magnitude cannot be left in the hands of inexperienced people.

79

ON ADVISORS

When I question the suitability of inexperienced people at the helm of highly technical institutions or companies, someone always attacks me, arguing that that is what consultants are for (of course, no engineer has ever given me this excuse).

I usually respond to the above with the following anecdote, which just as well may have happened. Hernán Cortés, Cuauhtémoc and Malinche are in the jungle in Honduras. Cortés says to the latter: «Let him tell you where the treasure is.» Malinche obeys and questions Cuauhtémoc, who replies: «I will never tell you.» Cortés then orders his feet to be burned and asks Malinche to repeat the question. Cuauhtémoc again refuses to disclose the location. Cortés orders the fire to be stoked and asks Malinche to ask him again.

Cuauhtémoc, with indescribable pain, tells Malinche: «It is buried in the grotto, by the ahuehuete, behind the main pyramid.» Malinche immediately informed Cortés: «He is not going to tell you.»

This is what happens when consultants do not know the subject matter: you are in their hands.

80

ALWAYS BE SOCIALLY RESPONSIBLE

As engineers, we need to maximize performance at the lowest cost. Failure to begin this way will lead to ineffective solutions with questionable efficiency.

However, this does not imply that once we find the solution, we should not see its consequences and repercussions for both society and the environment.

I remember when I was at university in the Underground Systems Course with Professor Guillermo Garibay Borrego, an expert on the subject and responsible for this area in the CFE. He asked us about the benefits of installing subway conductors in distribution systems and I was quick to answer that it was aesthetic. Needless to say, I was almost crucified. The typical engineer's answer would have been higher current density, lower failures and dozens of other factors. All very valid, but I remember that I mentioned that a transmission tower cannot be installed in front of the Palace of Fine Arts. Time passed and now the Mexican Magic Towns program incorporates, as a fundamental piece, burying the electrical system.

Another example is not designing products with their full life cycle in mind: how will it eventually be disposed of, is there anything that can be used at the end of its useful life? If we do not consider this, it will certainly be a product that will present problems for the owner in the future. We will see this with electric vehicles. To think in this way is to go beyond legislation. Just as with standards, it is the *mínimum minimorum.* So, take a long-term view and always consider the impact that something will have.

81

COMMERCIAL PRESSURES

As engineers, our loyalty is to technology and to the safe and proper functioning of products and services.

We live in a world where competition is increasingly fierce; however, that is no excuse for compromising our work ethic.

We must learn to say «no» and to refuse to do something that we know will result in problems sooner or later.

It is preferable to resign and even denounce something, rather than compromising our moral integrity and affecting the well-being of others.

There are plenty of excuses, but it is the facts that matter.

82

DO NOT GIVE SPACE FOR MISUNDERSTANDINGS

As engineers, we must be clear. Our position, especially in matters of safety, cannot be in any doubt.

This implies confronting equals and superiors when it would be easier and more comfortable to keep your mouth shut. It is sad to see that some engineers are ignored by arrogant and ignorant bosses.

Accepting a situation like this may set in motion the decline of our quality of life, and it may be very difficult to recover: every centimeter you cede will take a kilometer to reclaim.

If you make your position clear from the beginning, you will gain the respect of others. Perhaps some readers will ask «what if I get fired?» Maybe that would be a blessing: «every cloud has a silver lining.»

Do not lose your fortitude, as this quality allows you to grow as a person and ennobles the profession.

83

PREMEDITATIO MALORUM

This Latin expression should be the cornerstone of engineering. It means «premeditation of adversity,» that is, foreseeing things that can go wrong or thinking about what can go wrong. Seneca referred to this a lot: think of the bad things that can happen to you.

Force yourself to reflect on possible scenarios and look for the edges that can derail a project. Naturally, this may be something you will want to avoid as it is not pleasant to think about scenarios that are contrary to our interests. But like it or not, you will need to face them in the end.

If you do not prepare for the worst, you will simply not be ready. What could go wrong with my design? Is an underground substation in a rainy area a good idea? Did I check the seismic activity? What was the strongest wind in the area in the last hundred years? Do the personnel have the necessary training and experience? Thinking about these types of considerations can free you up because you have analyzed the possibilities with a cool head.

This is not meant to make you paranoid or permanently worried. On the contrary, it is for you to have peace of mind and to know how to react.

Normally, when things go wrong due to a lack of analysis of the different scenarios that may occur, we blame it on luck, which is neither professional nor valid. Luck favors those who prepare. As an engineer, you must think about what can go wrong and prevent it.

Another benefit of considering everything that could happen is that if you raise these concerns with your superiors, you demonstrate that you are disciplined and covering potential risks. Go beyond the day-to-day.

In addition, this discipline is very useful in everyday life. What happens if I miss my flight? Does my insurance cover me? How likely am I to get cancer with my family history? Like these, there are dozens of other examples that we can and should consider.

84

APPRECIATE WHAT ENGINEERING HAS PROVIDED TO HUMAN PROGRESS

Development over the last 150 years far exceeds what humankind achieved in 6,000 years, taking the invention of writing as a starting point.

The very speed of change has made us largely fail to appreciate what has been developed thanks to engineering. Even at the risk of upsetting my medical acquaintances and friends, I would argue that a physician from the beginning of the 20th century could make a more accurate diagnosis than today's doctors. Why? Because of their «clinical eye.» Physicians may boast about progress in medicine when the merit actually belongs to all the developments made by engineers.

This is often overlooked, but it is important to keep in mind, as if you do not believe in what you are worth, even less so will those around you. Of course, we should support the progress of humanity, but the benefits should be shared and we should ensure that there is a social purpose.

A hospital or a doctor will try anything to lower the cost of equipment to be purchased, but are personally offended should anyone ask them for a discount!

Value your talent and ability, do not let others reap what you have sown.

85

«TAKE OFF YOUR ENGINEERING HAT AND WEAR THE MANAGER'S» (NEVER FORGET YOUR RESPONSIBILITY)

The phrase above in quotation marks may seem innocuous, but it caused the explosion of the space shuttle Challenger.

The story goes like this: the head of the engineering department, who was checking for any problems that might arise together with his team, found that the low temperature on the day of the launch could cause a problem with the launch engine gaskets. When he suggested postponing the launch until he was sure that everything would go well, he received the suggestion from his boss to «take off his engineering hat and put on his manager's»: it was, as we know, a costly mistake.

Never, I repeat, never let a security issue get in the way of someone who is not a technician. To a greater or lesser extent, we all face these situations and must be prepared to say «no» and even (again) to resign or to denounce.

Our colleagues in Canada wear a steel ring as a badge; it is simple, without any engraving or embossing. The history of this is very illustrative; they wear it as a reminder of the steel bridge that crossed the St. Lawrence River in Quebec that collapsed due to an engineering error. More than seventy people died in the accident.

The ring on their finger is a permanent reminder, especially for new engineers, to not make the same mistake. What is your reminder of professional responsibility? In my case, the IEEE pin that always goes on my coat lapel.

86

AT THE START OF YOUR PROFESSIONAL LIFE CHOOSE A COMPANY FOCUSED ON ENGINEERING (AFTER THAT, TOO)

To convey this message in simple terms, let's start by dividing companies into those in which success is not measured by return on investment or profits - at least in the short term - but rather on quality (I take safety for granted) and then cost. On the other hand, there are those that always seek to maximize in the short term and as long as they comply with the «I got away with the minimum» they feel satisfied.

Undoubtedly, you will learn and enjoy more in the first ones, focused on quality; moreover, according to detailed studies, they are the most successful companies in the long term.

Make no mistake in your choice: give top priority to engineering.

When a man knows not which port he sails, no wind is favorable.

Lucius Annaeus Seneca

87

SURPRISES OF PROFESSIONAL LIFE

When studying at university, we may have thought that that was real life, that decisions made by a designer are based on elaborate calculations. Once we joined the working world, we realized that those decisions are made with such speed that there is obviously no calculation involved. What we witnessed was the accumulated knowledge and its true application. It is this that is at the heart of engineering.

Years of work enable us to perform these calculations without a major problem and, moreover, considering the real application. Therefore, it is important to ask colleagues how they came to their conclusions.

One of the things that bothers me most is when an engineer has to use a calculator to solve even the simplest of problems.

Some years ago, at a conference, I gave a stated value of a breaker as 600 amperes. One of the attendees raised his hand about thirty seconds later and commented: «The correct value is 587.70 amperes.» This happened when we were choosing a switch for a capacitor bank that had to be 1.35 times the rated current.

The difference would not influence the result, but would lead to errors. The «engineer» who made the comment insisted that using a 600-ampere breaker was not adequate, while using the «correct» one (not commercially available) would result in a cost increase without any benefit.

Therefore, listen to experience once you have acquired it, and not only to the calculator or an internet consultation.

88

PLANNING AND RISK

Correct planning allows you to achieve your objectives in a better way, but you should be careful not to fall into useless exercises that only serve to feed your ego.

Remember Immanuel Kant's quote, «Plan-making is often a luxurious and pretentious mental occupation, whereby the reputation of a creative genius is acquired by demanding what one cannot achieve oneself, censuring what one cannot improve, and proposing what one does not know where to find.»

In short, plans must be concise, achievable and have a completion date. The rest is mental verbiage.

As engineers, we have to make decisions that inevitably involve risk to users and to society. Determining what level of uncertainty is acceptable is vital and, as a rule of thumb, what the law says is only the *mínimum minimorum*, but by no means what should be acceptable.

Many of the most notable tragedies could have been avoided at little additional cost. For example, in the Fukushima nuclear accident in Japan, building a plant facing the Pacific, where a tsunami was possible, was considered to have a much greater risk. This implied two solutions. The first was to install the plant, like all others in the country, facing the China Sea; the second was that it should have a breakwater and redundant systems for cooling the reactors. Although more robust systems were considered in view of the risk involved in the location, they were, as we know, insufficient.

The force of nature is a factor that should be treated with great respect, so precautions are never too much. As my grandmother used to say: «It is better to err on the side of excess than to err on the side of defect».

89

THE ENVIRONMENT AND WHERE TO WORK

Day by day, regulations are tightening and, more importantly, public attention on the issue of environmental protection is forcing special care. What was once considered a minor issue is now becoming a major one.

As with security, matters related to interaction with people should not be seen as a one-off but rather, thought of for the long term. It is essential to consider the life cycle of a product, and this is far from simple.

In short, I believe that when designing we must consider: 1) the objective and scope: the product or process must be defined according to the limits of its use and its effects on the environment; 2) the requirements for its use: analyze the resources it will demand -energy, water, materials- and measure their effect; 3) an impact assessment: its effect on society and the environment; and 4) a diagnosis: with the data collected, decide on the best alternative or rethink the problem.

This reflection may lead you to consider what kind of company you want to work for. As a junior engineer, it is difficult to change a company's philosophy to take a greater interest in the environment. For the environment, therefore, it is important that you take into account the *modus operandi* of each institution.

In general, there are three types: 1) Those that are below the minimum - they pretend to comply, but avoid reaching the minimum and do not invest in this issue; 2) Those that fully comply - they respect laws and regulations, but see these as an unnecessary burden; and 3) Those with a progressive philosophy - there is conviction from the top management and they get really involved.

Of course, among the above options, the third is where we would like to work. It should not be forgotten that companies with this attitude usually have the same philosophy in other areas (respect for employees, ethical treatment of their suppliers and long-term vision). You have to know how to choose.

90
ON SIMULATIONS AND SCIENCE

One way to avoid going down a path that ends in a dead end is to carry out simulations that allow us to know if our effort will lead to a chance of success.

Start with a simple simulation and, if the result is encouraging or promising, refine it and adjust the model (we have a general tendency to try to create overly complex models).

Remember Voltaire: «Perfect is the enemy of the good.» Do not get lost in detail. This is the difference between a good engineer and a mediocre one. Many of the solutions we propose are not practical, and in many cases, in striving to be perfect, a project either goes over budget or becomes completely impractical.

It is worth remembering that we are engineers, and it is natural for newcomers to confuse science with engineering. The two complement each other, but we have to distinguish them well. The simplest way is to be clear that in science, if you know what you are doing, you should not be doing it. In engineering, if you do not know what you are doing, it is best to stop. This is the only way to avoid wasting time and resources.

91

WE LIVE IN A HIVE

«What is not useful for the hive is not useful for the bee.» This phrase, taken from the *Meditations* of Marcus Aurelius, reminds us of what we know but forget or, perhaps more accurately, despise.

In the field of engineering this is even more important, since our work is in real time. Everything is interconnected, so a failure in the hive affects a large number of bees.

Do not lose sight of the big picture in the actions you take, and always consider their short-, medium-, and long-term consequences.

An improperly calibrated protection scheme or the failure to check if batteries are working properly can cause problems for an entire population. For example, in an industrial park in Ciudad Juárez, Chihuahua, a transformer explosion caused the pumping system in the southern part of the city to stop. More than 300,000 inhabitants suffered from a lack of water for more than eight hours. A small detail that affected the entire hive.

92

BE THE BEST ENGINEER THAT YOU CAN... AND DO NOT FALL INTO SWEET INERTIA

Although we have already covered the importance of believing that you can do something, it never hurts to emphasize it.

Although «never» is a prohibited word, we can almost certainly say that we will not surpass Tesla, that we cannot compare ourselves to him. Nevertheless, this does not imply that we do not have ambitious goals that will allow us to develop as people. Shooting high to fall in the middle always works. Of course, it is useful to put it in writing and look for ways to strengthen our goals.

As engineers, we understand well the concept of inertia, but Tacitus, the Latin poet of the first century, coined the term *inertiae dulcedo.* He was referring to the comfortable reluctance to take risks; to the temptation to not transgress the scale of values, even when we know it is incorrect, in order to avoid conflict or concerns; to the cowardice and lack of character that shackles us.

Why does it exist? Because self-censorship is always simpler than censorship by others. We experience it every day when we decide to turn our heads and ignore what we know is wrong. From inadequate safety conditions in our plant («nothing ever happens») to witnessing injustice.

We engineers cannot afford this attitude, which unfortunately increases every day under the cover of «political correctness» – a euphemism for cowardice.

Do not let this stupor turn into inaction. Only then will you be the best engineer you can be.

93

THINK BEFORE USING THE INTERNET

Access to smartphones often causes us to consult the internet before thinking. In the 2nd century B.C., Eratosthenes was able to calculate the circumference of the earth during the summer solstice with only a 1.16% error! For this he used the shadows and knowledge of distances with accuracy.

It must also be said that he had all the knowledge at his disposal, since he was the director of the Library of Alexandria, a precinct that, prior to the fire, housed all the available knowledge.

So next time you have a question, work out the answer first and do not go for the easy solution. More brains, less Google.

94

SCIENCE FICTION

Engineers are naturally drawn to the possibilities offered by technologies that are possible but that do not yet exist, either in the form of literature, from Jules Verne to Carl Sagan and Aldous Huxley or, today, with *Star Wars* and *Star Trek* movies (my favorite).

Since I am a trekkie I am going to concentrate on the latter, in particular *The Next Generation* with Captain Jean-Luc Picard and his companions.

This installment presents advances in cooking equipment (food replicators), which we have already approached with the Thermomix; giving instructions by voice to a computer («Alexa: what is the temperature?») and the use of new materials that were considered impossible.

The Enterprise flies faster than light, at warp speed. But considering Einstein's formula ($E=mc^2$), this is not possible, it escapes explanation. It all started when Jean-Luc Picard raised his hand and commanded: «Warp 8, engage». The spacecraft became a wake and exceeded the speed of light.

It turns out that a Mexican, Miguel Alcubierre, a hardcore trekkie, discovered that traveling at warp speeds is, in fact, possible. Without the inspiration of the series, or perhaps fanaticism, the development of this technology would probably not have been attempted or even considered. Dreaming always pays off as long as you do not get lost in pipe dreams. As Captain Picard once said: «Things are impossible until they are no longer impossible.»

95

COMPETING WITH GREAT RIVALS MAKES YOU BETTER

In general, we think that little can be learned in a small or medium-sized company, but this is a limited point of view.

If you work in such a company, you are faced daily with those who are better than you and who may have greater resources. This is something you should embrace as an opportunity.

When I started Inelap - a Mexican company dedicated to the manufacture and marketing of electrical equipment to improve the quality of electricity supply - we were competing against the big names in the industry. This forced me to get up earlier every day, study harder and try to be closer to our customers.

There was no greater incentive than to beat them on their own turf, especially as Mexicans: we learned not to bow our head.

You cannot be a better engineer without going up against the best. Sooner or later, you will get the opportunity to be the role model.

This is not to say that learning in a large company is not possible, and perhaps a very good first step, but in the long run you will look more inward than in the direction of market opportunities.

96

UNDERSTAND AND RESPECT THE FORCE OF NATURE

As engineers, we pride ourselves on having transformed the world more in one century than it had in five millennia. However, if we take all the annual energy produced by electrical systems, add the energy used in transportation, industry, commerce, and households, and convert it into a unit, it would still only be equivalent to that released by a hurricane from onset through to tropical storm. A monsoon in Asia releases even more: ten units and the annual global rainfall of several thousand units.

This gives us an idea of our relative insignificance and puts into perspective what nature can generate. It always surprises us.

But, on the other hand, we have been able to control water, as evidenced by the construction of the Suez and Panama canals, engineering projects proposed by the Frenchman Ferdinand de Lesseps. So, the lesson is to maintain humility in the face of nature, and assume that there may be errors in your calculations.

97

«I WILL REPEAT MYSELF UNTIL YOU UNDERSTAND ME»

This quote from Voltaire (again) is highly relevant for us engineers. It is imperative to make sure that your message has been received, analyzed and, crucially, understood. To do so, you should ask for it to be repeated back to you and confirm that it has been duly processed.

Dozens of accidents of all kinds come from assuming that a message or instruction has no chance of being misinterpreted. Only with Prussian discipline and without assumptions is that possible.

Remember the game of the broken telephone? When a message is passed from one person to another, without being questioned, it is totally distorted by the time it reaches the last participant. So, do not assume that your message has been understood, as you could waste valuable time. Of course, it will be much clearer with less chance of error if it can be put in writing.

98

THE CONFIDENCE THAT YOU PROJECT

«When a train goes through a tunnel and it gets dark, you don't throw away the ticket and jump off. You sit still and trust the engineer,» said Corrie ten Boom. This quote aptly sums up what our profession is all about. In fact, we are so efficient that we are not thanked for it. It is taken for granted.

An airplane taking off and landing is a marvel that we take for granted. The same is true when we flip a switch and have light. The greatest recognition may come from family or friends who are not engineers. It is even hard for them to grasp the magnitude of what we have built.

We are only a small part of this huge engineering system, so you have an obligation not to disappoint or betray the trust placed in you.

99

ON THE DEFENSE OF THE PROFESSION AND THE SECTOR

Some years ago, I wrote an article for *Energía Hoy* magazine entitled «Why not to be an Engineer» (https://goo.su/YjgL), a theme I continued a few months later. (https://goo.su/GOay). Of course, the conclusion was obvious: it pays to be one, but we engineers are our own worst enemies.

I mentioned that there are engineers who ask their colleagues to work for free with the false promise that the loss can be «recovered» through potential future sales; but this is not true, as an engineer who engages in this sort of behavior rarely has the respect of management nor the authorization to place a purchase order. In fact, they are normally considered a nuisance or a necessary evil.

It is unprofessional to ask someone to work without receiving the corresponding reward. Question: would these pseudo-engineers have the guts to ask the same of a doctor or a lawyer?

If it went no further, perhaps it would not be so serious, but the situation permeates, and we end up being paid much less than other professions, such as administrators or lawyers. Maybe it is because we deserve it.

It is an insult to the profession and resembles the practice of *harakiri*. We complain about meager incomes, but we encourage it. Without realizing it, we fall into a downward spiral, and it is hard to get back to where we started.

The Director of Corporate Purchasing for an auto parts company recently asked me to itemize my services of a network code study. I refused and he replied, «We give it to our customers,» which is certainly true, auto manufacturers have enormous buying power. I then replied that if he gave me thousands of network code studies I would do it, but not for a one-off.

At another company I was asked to be responsible for reviewing work that they would implement. I presented a quotation, and, as expected, they asked me how many hours it would take me to do it. I answered six and, immediately, the person who attended me commented: «That is more expensive than the chief external lawyer who handles corporate matters. How do you justify it?» My answer was succinct: «I am not going to justify it, but if you present me weighty arguments that justifies why a lawyer should earn more per hour than an engineer, I will do it for free.»

As expected, this person was upset and we did not come to any agreement. However, out of respect for our profession, I refuse to let the standard be the income of a lawyer. As a corollary, I will comment that one of the client's engineers who was present called me afterwards to congratulate me.

The story ended well anyway. They hired an engineering «firm» that delivered something unusable and asked us to correct it. In the end, they ended up paying almost the same and lost two months.

Again: do not let the profession degrade.

100

HOW MANY ENGINEERS ARE THERE IN MEXICO? (THE INCLUSION OF FEMALE TALENT)

We are currently almost two million engineers in Mexico and 120,000 are added every year. I consider that the number makes us a cornerstone of the country's growth; for me this is very important and is one of the reasons for writing this book.

We still have a long way to go in certain engineering disciplines, it is true, and one of the great challenges is the inclusion of more female talent. We have made progress in this regard because, at least in areas such as chemistry and the environment, women already make up the majority.

The following table shows the percentages of men and women in the different engineering degrees (for ease of reading I have rounded off the figures and percentages):

Engineering	Graduates (total in thousands)	% Men	% Women
Industrial	520	65	35
Civil	310	87	13
Mechanics	280	89	11
Chemistry	270	41	59
Electronics	260	92	8
Electric	130	91	9
Automotive	60	99	1
Environmental	24	42	58

Source: Mexican Institute for Competitiveness (IMCO, 2023).

The road is long, but we need to encourage women's participation in engineering even further.

Museums that every engineer should visit before they die

- Deutsches Museum von Meisterwerken der Naturwissenschaft und Technik (Munich, Germany)
- Musée des Arts et Métiers (Paris, France)
- Science Museum, Universum, UNAM (Mexico)
- Museo delle Scienze, MUSE, (Trento, Italy)
- Museo Galileo [former Istituto e Museo di Storia della Scienza] (Florence, Italy)
- Nikola Tesla Museum (Belgrade, Serbia)
- Technological Museum of the Federal Electricity Commission (Mexico)*
- Museu da Electricidad (Lisbon, Portugal)
- Museum of Innovation and Science, miSci, [IEEE Archives] (New York, United States)

---

*This used to be the second most visited in the country. Unfortunately, it closed in 2015 and there seems to be no plan to reopen it.

- Smithsonian Institution (Washington, DC, United States)
- SPARK Museum of Electrical Invention (Washington, United States)
- The Energy Museum (Berlin, Germany)
- The National Museum of Emerging Science and Innovation, Mirakain (Tokyo, Japan)
- The National Science Museum (Seoul, South Korea)

Epilogue

Despite the many activities and concerns you may have, you have reached the end of the book, and I am grateful that you finished it. I know that I have an advantage, given that I wrote the book and so the concepts are clearer in my mind.

Nevertheless, you should constantly practice the tips you have just read so that they become well engrained and a natural habit. It is claimed, quite rightly, that to master something you have to practice it for 10,000 hours.

Do not stop and you will get there quicker than you think. It is not that hard. The journey starts with the first step, but I guarantee you that it is a great journey and, when you move forward, results come in spades.

In my case, the greatest recognitions have been the least expected, while when we are about to finish a project or a study plan, to some degree we anticipate the outcome.

At Expo Eléctrica 2021, I was attending clients at the PQ Barcón booth, when a colleague told me that a young woman wanted to meet me and have me sign our general catalog. As she was an admirer I immediately agreed. Her name is Arely Hernandez. After the introduction, and while I was signing the catalog, I asked her about her studies. She answered that she was an engineer. I put down the catalog and picked up the book *Power Quality. Power Factor and*

Harmonic Filtering, which I had co-authored, and wrote a dedication to her.

Her eyes immediately misted over, and, in a natural movement, she placed her hands on her chest, as if in prayer. Needless to say, it was by far one of my best experiences as a professional.

Two days later I was interviewed on *Mundo Sustentable*, a program that airs every Saturday at 8 a.m. on Imagen Radio, both on TV and radio, and before starting I shared this anecdote. They mentioned it at the beginning, giving me, in passing, the nickname of rockstar.

It takes time, but with certainty I can say that decades of effort materialized in that instant.

If you have any comments or questions I would love to hear from you. Feel free to contact me at the following email address:

santiago.barcon@pqbarcon.com

Bibliography

Aurelio, M. (2022). *Meditaciones*. Penguin Libros.

Bartlett, J. (2012). *Bartlett's Familiar Quotations*. Little Brown and Company.

Berlin, I. (2013). *Against the Current: Essays in the History of Ideas*. Princeton University Press.

Bloetscher, F. (2021). *Ethics and Responsibilities of Engineers*. J. Ross Publishing.

Blount, J. (2018). *The Ultimate Guide for Mastering the Art and Science for Getting Past No*. Wiley.

Carbajal, C. S., & Chávez, E. (2008). *Ética para ingenieros*. Grupo Editorial Patria.

da Empoli, G. (2019). *Les ingénieurs du chaos*. Gallimard.

Damasio, A. (2005). *Descartes' Error: Emotion, Reason, and the Human Brain*. Penguin Books.

Davis, M. (1998). *Thinking Like an Engineer*. Oxford University Press.

Didier, C. (2013). *Les ingénieurs, les risques technologiques et l'éthique Professionnelle*. Hal Open Science. https://shs.hal.science/halshs-00779968/document.

Dörner, D. (1989). *The Logic of Failure*. Basic Books.

Dobelli, R. (2016). *El arte de pensar claramente*. Ediciones B.

Energy Institute. *Code of Professional Conduct*. (2018). https://www.energyinst.org/data/assets/pdf_file/0014/541022/The-EI-Code-of-Professional-Conduct-january-2018.pdf.

Florman, S. (1976). *The Existential Pleasures of Engineering*. St. Martin's Griffin.

Florman, S. (1987). *The Civilized Engineer.* St. Martin's Griffin.

Geisaler, P. (2015). *The Power of Writing Well: The Thoughtful Leader's Model for Business and Technical Communications.* The Expressive Press.

Govindarajan, M., Natarajan, S., & Senthilkumar, V. S. (2004). *Engineering Ethics (Includes Human Values).* PHI Learning.

Graupera, V. (2018). *Time Management for Engineering Mannagers.* Self-Publishing.

Hamming, R., & Victor, B. (2010). *The Art of Doing Science and Engineering: Learning to Learn.* Stripe Press.

Harris, C., Pritchard, M., Rabins, M., James, R., & Englehardt, E. (2009). *Concepts and Cases: Engineering Ethics.* Cengage.

IESF Société des Ingénieurs et Scientifiques de France (n. d.). *Responsabilité éthique de l'ingénieur dans les systèmes complexes.* https://www.iesf.fr/offres/doc_inline_src/752/CHAPITRE_B3_Responsabilite_ethique.pdf.

Ingénieurs Canada. *Code of Ethics.* https://engineerscanada.ca/guidelines-and-papers/public-guideline-on-the-code-of-ethics#.

Institute of Electrical and Electronics Engineers (IEEE). *Code of Ethics.* https://www.ieee.org/about/corporate/governance/p7-8.html

Instituto Mexicano de la Competitividad (IMCO). (2023). *Compara carreras.* https://imco.org.mx/compara-carreras-2023/

Kaplan, B., & Manchester, J. (2018). *The Power of Vulnerability.* Greenleaf Book Group Press.

Lapham, L. (Ed.) (2008). *Ways of Learning. Lapham's Quarterly, Fall Issue.* American Agora Foundation.

Layton, E. (1971). *The Revolt of the Engineers.* Johns Hopkins University Press.

Lladó, M. (2020). *Enamorarse del futuro.* Plataforma Empresa.

Maalouf, A. (2021). *Adrift: How Our World Lost Its Way.* World Editions.

Montaigne, M. (2007). *Les Essais.* La Pléiade, Gallimard.

Munson, T. (2018). *People Skills for Engineers.* Independently Published.

Parrish, S., & Beaubien, R. (2019). *The Great Mental Models.* Farnam Street.

Petroski, H. (1982). *To Engineer is Human.* Vintage Books.

Royal Academy of Engineering (2022). *Statement of Ethical Principles.* https://www.engc.org.uk/ethics

Ruiz, J. C. (2020). *El arte de pensar.* Almuzara.

Savater, F. (2021). *Solo integral. Una vuelta de tuerca a sus mejores ideas.* Ariel.

Sinek, S. (2019). *The Infinite Game.* Portfolio.

Slocum, S. (2009). *She Engineers.* Engineers Rising.

Spektrum, & Gast, R. (July 13, 2021). Star Trek: Warp Drive Leads to New Physics. *Scientific American.* https://goo.su/nnhH0dz.

Vincenti, W. (1990). *What Engineers Know and How They Know It.* Johns Hopkins University Press.

Whitbeck, C. (2011). *Ethics in Engineering Practice and Research.* Cambridge University Press.

32 years

Still much to do

LID

- 1993 Madrid
- 2008 Mexico City
- 2010 London
- 2011 Buenos Aires
- 2012 Bogota
- 2014 Shanghai

www.ingramcontent.com/pod-product-compliance
Lightning Source LLC
Chambersburg PA
CBHW071341230926
51846CB00002B/409
* 9 7 8 6 0 7 2 6 9 8 8 3 3 *